FERMENTED FOODS AND BEVERAGES IN A GLOBAL AGE

FERMENTED AND DISTILLED ALCOHOLIC BEVERAGES

A TECHNOLOGICAL, CHEMICAL AND SENSORY OVERVIEW

RED WINES

FERMENTED FOODS AND BEVERAGES IN A GLOBAL AGE

Additional books and e-books in this series can be found on Nova's website under the Series tab.

FERMENTED FOODS AND BEVERAGES IN A GLOBAL AGE

FERMENTED AND DISTILLED ALCOHOLIC BEVERAGES

A TECHNOLOGICAL, CHEMICAL AND SENSORY OVERVIEW

RED WINES

M. BONATTO MACHADO DE CASTILHOS
VANILDO LUIZ DEL BIANCHI
AND
VITOR MANFROI
EDITORS

Copyright © 2021 by Nova Science Publishers, Inc.

All rights reserved. No part of this book may be reproduced, stored in a retrieval system or transmitted in any form or by any means: electronic, electrostatic, magnetic, tape, mechanical photocopying, recording or otherwise without the written permission of the Publisher.

We have partnered with Copyright Clearance Center to make it easy for you to obtain permissions to reuse content from this publication. Simply navigate to this publication's page on Nova's website and locate the "Get Permission" button below the title description. This button is linked directly to the title's permission page on copyright.com. Alternatively, you can visit copyright.com and search by title, ISBN, or ISSN.

For further questions about using the service on copyright.com, please contact:
Copyright Clearance Center
Phone: +1-(978) 750-8400   Fax: +1-(978) 750-4470   E-mail: info@copyright.com.

NOTICE TO THE READER

The Publisher has taken reasonable care in the preparation of this book, but makes no expressed or implied warranty of any kind and assumes no responsibility for any errors or omissions. No liability is assumed for incidental or consequential damages in connection with or arising out of information contained in this book. The Publisher shall not be liable for any special, consequential, or exemplary damages resulting, in whole or in part, from the readers' use of, or reliance upon, this material. Any parts of this book based on government reports are so indicated and copyright is claimed for those parts to the extent applicable to compilations of such works.

Independent verification should be sought for any data, advice or recommendations contained in this book. In addition, no responsibility is assumed by the Publisher for any injury and/or damage to persons or property arising from any methods, products, instructions, ideas or otherwise contained in this publication.

This publication is designed to provide accurate and authoritative information with regard to the subject matter covered herein. It is sold with the clear understanding that the Publisher is not engaged in rendering legal or any other professional services. If legal or any other expert assistance is required, the services of a competent person should be sought. FROM A DECLARATION OF PARTICIPANTS JOINTLY ADOPTED BY A COMMITTEE OF THE AMERICAN BAR ASSOCIATION AND A COMMITTEE OF PUBLISHERS.

Additional color graphics may be available in the e-book version of this book.

Library of Congress Cataloging-in-Publication Data

Names: Castilhos, Maurício Bonatto Machado de, editor. | Bianchi, Vanildo
   Luiz del, editor. | Manfroi, Vitor, editor.
Title: Fermented and distilled alcoholic beverages--a technological,
   chemical and sensory overview. Red wines / Maurício Bonatto Machado de
   Castilhos (editor), Universidade do Estado de Minas Gerais, Minas
   Gerais, Brazil, Vanildo Luiz Del Bianchi (editor), Vitor Manfroi
   (editor).
Description: New York : Nova Science Publishers, [2020] | Series: Fermented
   foods and beverages in a global age | Includes bibliographical
   references and index. |
Identifiers: LCCN 2020055060 (print) | LCCN 2020055061 (ebook) | ISBN
   9781536189858 (softcover) | ISBN 9781536190526 (adobe pdf)
Subjects: LCSH: Red wines. | Wine and winemaking--Chemistry. | Wine--Flavor
   and odor.
Classification: LCC TP548.6.R43 F47 2020 (print) | LCC TP548.6.R43
   (ebook) | DDC 663/.223--dc23
LC record available at https://lccn.loc.gov/2020055060
LC ebook record available at https://lccn.loc.gov/2020055061

Published by Nova Science Publishers, Inc. † New York

Contents

Preface		vii
Chapter 1	American Grape Wines: The *Vitis labrusca* X-Factor *Maurício Bonatto Machado de Castilhos,* *Lia Lúcia Sabino,* *Jorge Roberto dos Santos Júnior* *and Vanildo Luiz Del Bianchi*	1
Chapter 2	Phenolic Composition of Bordeaux Cabernet Sauvignon and Merlot Grapes and Wines *Kleopatra Chira, Maria Reyes González-Centeno,* *Michael Jourdes and Pierre-Louis Teissedre*	21
Chapter 3	Malolactic Fermentation of Tempranillo Wines: Effects on Chemical Composition and Sensory Quality *Pedro Miguel Izquierdo-Cañas,* *Sergio Gómez-Alonso, Esteban García-Romero,* *Gustavo Cordero-Bueso and* *María de los Llanos Palop-Herreros*	53

Chapter 4	Red Wines: Carmenère *Carolina Pavez, Philippo Pszczólkowski,* *Natalia Brossard and Edmundo Bordeu*	79
Chapter 5	Touriga Nacional Red Grape Variety: Phenolic and Aroma Composition and Winemaking Technology *Fernanda Cosme, Luís Filipe-Ribeiro* *and Fernando Milheiro Nunes*	115
Chapter 6	Tannat Wine: Characteristics and Key Stages in Its Production *Laura Fariña, Karina Medina, Valentina Martín,* *Francisco Carrau, Eduardo Dellacassa* *and Eduardo Boido*	163
Chapter 7	Syrah (*Vitis vinifera* L.) Wines in Brazil *Juliane Barreto de Oliveira* *and Giuliano Elias Pereira*	197

About the Editors   227

Index   231

PREFACE

Wine is considered one of the most complex alcoholic beverages in the world since it presents several compounds that play a relevant role in wine quality. Generally, it is possible to consider that wine is a mixture of water and ethanol; however, if we consider this fact as a feasible definition, all the wines produced worldwide would present similar chemical composition and sensory profiles. We all know that wine cannot be defined as a simple mixture of water and ethanol, and the high complexity of the wine is determined by the presence of minor compounds, primarily the phenolic and volatile compounds.

All these minor compounds are responsible for wine color, aroma, body, flavor, and their importance for wine quality is undoubtful. These minor compounds are synthesized by several biochemical reactions such as alcoholic fermentation and malolactic fermentation, also can be derived from the winemaking procedure and its respective variations, and can be obtained naturally from the grape cultivar, promoting a varietal character for wine, a feature that is highly sought by the wineries. The search for the discovery of wine with unique sensory features is directly related to the presence of these minority compounds.

Also, the phenolic compounds existent in red wines, such as anthocyanins, flavonols, flavan-3-ols, phenolic acids, and stilbenes are responsible for promoting one of the principal wine properties: the antioxidant effect. Red wines, especially, have in their composition, different forms of anthocyanins, which play an important role in the red wine color. The anthocyanins, with all the rest of the phenolic compounds, especially the compounds of the stilbene class, promote the high antioxidant capacity for wine, showing the wine potential as a beverage that promotes health and avoids degenerative diseases.

The importance of wine complexity for the tasters also resides on its aroma profile. The minor volatile compounds, which are derived from the grape and the wine technology, are responsible for different aromatic notes, such as fruity and woody ones, as well as more complex aromas determined by some specific chemical compounds such as volatile phenols and lactones that promote a coconut-like, milky, and fruity aroma for wine.

In this volume, from the book entitled *Fermented and Distilled Alcoholic Beverages: A Technological, Chemical and Sensory Overview*, we invite the reader to have a pleasure reading regarding red wines produced all over the world. This volume presents seven chapters as follows: wines produced from American grapes, authored by Brazilian researchers (Chapter 1), wines from Bordeaux region, authored by French researchers (Chapter 2), Tempranillo wines authored by Spanish wine researchers (Chapter 3), Carmenère wines authored by Chilean specialists (Chapter 4), Touriga Nacional grapes and wines authored by Portuguese researchers (Chapter 5), Tannat wines authored by Uruguayan authors (Chapter 6), and Syrah wines authored by Brazilian researchers (Chapter 7). All the chapters present useful information about the winemaking technology and its variations, as well as relevant information about the chemical and sensory profiles of the wines produced from different regions of the world. I hope that the reader has a pleasant time reading all the information in this volume. Enjoy it!

Maurício Bonatto Machado de Castilhos
Editor
Exact Sciences and Earth Department
Minas Gerais State University, Frutal, Minas Gerais, Brazil

In: Fermented and Distilled
Editors: M. B. M. de Castilhos et al.
ISBN: 978-1-53618-985-8
© 2021 Nova Science Publishers, Inc.

Chapter 1

AMERICAN GRAPE WINES: THE *VITIS LABRUSCA* X-FACTOR

Maurício Bonatto Machado de Castilhos[1,*], *Lia Lúcia Sabino*[1], *Jorge Roberto dos Santos Júnior*[2] *and Vanildo Luiz Del Bianchi*[2]

[1]Exact Sciences and Earth Department, Minas Gerais State University, Frutal, Minas Gerais, Brazil

[2]Food Technology and Engineering Department, São Paulo State University, São José do Rio Preto, São Paulo, Brazil

ABSTRACT

Wine is one of the most famous beverages in the world, and the principal wine producers worldwide are European countries, which are known as their high quality wines produced from *Vitis vinifera* grapes. In contrast to this scenario, Brazil is one of the countries that present a production of *Vitis vinifera* wines; however, the production of wines produced from *Vitis labrusca* grapes stands out in comparison with the great wine producers. *Vitis labrusca* grapes are underestimated in

---

[*] Corresponding Author's E-mail: mauricio.castilhos@uemg.br.

comparison with the Vitis vinifera grapes since the wines produced from the American grapes are sweet and fruity, and these sensory features are not very appreciated by the wine experts and consumers. However, the American grapes and their hybrids have, in their composition, a great number of phenolic compounds accounting for high antioxidant capacity, color indexes, and stability. Therefore, the importance of these grapes and wines should be highlighted since these wines, depending on the grape cultivar, present features that make them unique.

Keywords: *Vitis labrusca*, red wine, winemaking, color, antioxidant capacity

INTRODUCTION

Wine is one of the most studied beverages worldwide, and several countries are responsible for delivering interesting researches, showing relevant results regarding wine chemical profile and its response in the sensory assessment, as well as alternative winemaking techniques to improve wine quality. The wine world production, in 2019, was 260 million of hectoliters, and among the major worldwide wine producers, Italy, France, and Spain lead the wine production with 47.5, 42.1, and 33.5 million of wine hectoliters. These three European countries are responsible for 47.3% of the whole world wine production. Brazil also presents its importance in the wine world scenario due to its wine production of 3.1 million of hectoliters in 2018, and prospection of 2.0 million of wine hectoliters in 2019 [1].

The wine from the major wine producers in the world usually is elaborated using *Vitis vinifera* grapes, which provide wines with varietal characteristics, producing balanced beverages with aging potential, i.e., wines that can be submitted to the maturation process performed in the bottle or oak casks. This procedure can enhance the wine sensory features, making the tannins smooth, improving wine aroma, and providing the exchange of the wood compounds for the wine. However, Brazil is one of the most producers in table wines, produced from *Vitis labrusca* grapes, which present low importance for the wine field since they are known as grapes that produce wines with low quality. In 2018, Brazil has produced 1.6 million grape tons, and 0.8 million tons were *Vitis labrusca* grapes, also known as table grapes, accounting for 50% of the Brazil total grape production [2].

Wine is considered a complex beverage, and its complexity is related to the high amount and variety of minor compounds that are responsible for its aroma and flavor. Among the primary compounds that present significant importance for the wine varietal aroma and flavor, the volatile and phenolic compounds are responsible for promoting the wine's unique features. In this context, several studies have been reported the use of alternative winemaking techniques to obtain a wine with a different chemical profile, which can be responsible for the crucial improvement in wine sensory profile.

The present chapter provides relevant information regarding the *Vitis labrusca* wine production, showing alternative technologies employed in American grapes winemaking, and presenting the chemical and sensory behavior of these wines when produced from traditional and alternative winemaking techniques.

WINEMAKING PROCEDURE

The traditional winemaking process presents several steps that are commonly applied in wineries worldwide; however, several works have been reported variation in the traditional winemaking process depending on the location. Among the variations observed, it is possible to notice the different concentrations of potassium metabisulfite used for alcoholic fermentation [3, 4]; the variation regarding the yeast concentration used for the alcoholic fermentation induction [5, 6]; the application (or not) of the malolactic fermentation promoting smoothness for wine, primarily due to a decrease on the wine acidity [6, 7, 8]; the use of malolactic fermentation in different forms: induced or spontaneous [4, 5]; the application of clarification agents [5, 9], among others.

One of the common practices employed in winemaking using *Vitis labrusca* grapes is the correction of the alcohol content using sucrose. This practice known as chaptalization [10] is allowed in some countries, and the Brazilian legislation suggests a maximum limit of 3% v/v ethanol correction [11]. The chaptalization practice is allowed for wines produced from *Vitis labrusca* grapes due to their low sugar content in their optimal maturity stage. The low sugar concentration of these grapes does not achieve the minimum alcohol content allowed by the legislation (8.6% ethanol v/v). This practice is avoided in several countries that are known for their relevant wine production using Vitis vinifera grapes since these grapes present in their composition a higher amount of sugar in their optimal stage of maturity and, therefore, they do not present this particularity [10].

Winemaking technology can be considered a complex technique since it involves several steps that occur simultaneously. As mentioned above, there is difficulty in standardizing the traditional winemaking process since some factors influence the way that the wine is produced as follows: wine and winery reputation, region, wine aging, color, chemical profile, and vintage. Also, some of the traditional methods, such as filtration, clarification, and tartrate stabilization, are not usually applied by some wineries worldwide [12].

The traditional winemaking procedure consists of destemming and crushing the grapes allowing the release of the grape must. The obtained must is the primary substrate for the alcoholic fermentation. Before the alcoholic fermentation takes place, the grape must is treated with sulfur dioxide (using the potassium metabisulfite form), which is responsible for providing an antioxidant and antimicrobial activity for wine must. The sulfur dioxide antimicrobial activity is selective, i.e., this property only affects microorganisms that are considered harmful for the wine process, such as some species of bacteria and fungus [10, 13]. There is a discussion about the sulfur dioxide concentration used for avoiding contamination since the concentration applied should be directly dependent on the grape sanitary conditions, i.e., the high the grape sanitary conditions, the lower the sulfur dioxide concentration, and vice versa [10, 13].

Oliveira et al., (2019) [4] reported the use of 50 mg.L-1 of sulfur dioxide in Syrah wines produced in Brazil, Panceri et al., (2015) [5] reported the use of 100 mg.Kg-1 of sulfur dioxide in Cabernet Sauvignon and Merlot wines submitted to grape dehydration, De Castilhos et al., (2020) [14] and De Castilhos et al., (2019) [9] reported the use of 80 mg.L-1 of sulfur dioxide in BRS Rúbea and BRS Cora (*Vitis labrusca* hybrid grapes) red wines, and BRS Carmem and BRS Violeta (*Vitis labrusca* hybrid grapes) red wines, respectively; Lee et al., (2016) [8] also reported the use of 200 mg.L-1 of sulfur dioxide in Korean red wines produced from *Vitis labrusca* grape cultivars. These observed differences regarding the sulfur dioxide concentration are directly related to the grape sanitary conditions and the natural grape resistance against harmful microorganisms.

The alcoholic fermentation is a biochemical reaction that metabolizes sugars, primarily glucose and fructose, into ethanol and carbon dioxide. The Saccharomyces cerevisiae yeast specie is considered the primary microorganism responsible for the alcoholic fermentation since it presents high osmotolerance in comparison to the other yeasts [10, 13].

Other yeasts such as Hanseniaspora uvarum, Torulopsis bacillaris, and Kloeckera apiculata can participate in the alcoholic fermentation biochemical reactions; however, they usually act in the initial stages of this biochemical process. Depending on the winery and the vine management, wild yeasts can directly participate in the alcoholic fermentation, and some of them can cause negative effects on wine such as off-flavors and off-odors. Some of these wild yeasts can be exemplified by Kluyveromyces, Schizosaccharomyces, Zygosaccharomyces, and Brettanomyces genus.

Brettanomyces genus is a controversial yeast that causes strong discussion on the enology field since authors believe that this yeast produces volatile phenols chemical compounds that are responsible for the appearance of wine off-flavors and off-odors. The primary volatile compounds responsible for this negative effect are 4-ethyl phenol (4-EP), 4-ethyl guaiacol (4-EG), 4-vinylphenol (4-VP), 4-vinyl guaiacol (4-VG), producing a mousy, medicinal, wet wool, burnt plastic, or horse/sweat smells/scents for wine [15, 16]. In contrast, Crauwels et al., (2015) [17] reported that the Brettanomyces anomalus, also known as Brettanomyces claussenii or Dekkera anomala, has the potential to provide or enhance the floral and fruity wine features, making it more attractive to the consumer. In general, the concentration used for the Saccharomyces cerevisiae used in the inoculation onto the fermentative must is around 20 g.hL-1 [3, 4, 7, 8]; however, Panceri et al., (2015) [5] have reported the use of 25 g.hL-1 yeast concentration.

After the yeast inoculation, the grape pomace still in contact with the grape must, and the period of contact between them is known as maceration. During the maceration, the wine and pomace inside the fermentation flask are submitted to punching down and pumping over movements for better extraction of the color (anthocyanins presented into the grape skins) and for yeast growth enhancement, breaking the cap formed by the agglomeration of the solid parts inside the fermentation flask. During the alcoholic fermentation, the wine increases its ethanol content as a result of the carbohydrate metabolism by the yeasts, resulting in a decrease in wine density. Near to the end of the alcoholic fermentation, around 7 days for *Vitis labrusca* wines, the pomace is separated from the wine using a technique called dejuicing. The remaining pomace is then pressed, allowing for the release of nearly 10 to 15% of the juice [10, 12, 13, 18].

After the dejuicing, the wine is maintained at controlled temperature (around 22 to 23 °C) for 10 days followed by three racking procedures. The racking procedure consists of separating all the suspended solids into the wine solution to make it more transparent and clear. The first racking is carried out after the phenolic and protein stabilization, the second one after the malolactic fermentation, and the third one after the tartrate stabilization. After the first racking, the wine stays for more than 10 days, allowing the second fermentation known as malolactic fermentation. This fermentation can take place spontaneously or can be induced by the inoculation of acid lactic bacteria Oenococcus oeni, which causes the decarboxylation of the malic acid into lactic acid, promoting light and smooth acidity to the wine.

Thin Layer Chromatography (TLC) using a mixture of butanol, acetic acid solution at 50%, and bromophenol blue indicator [13] follows the end of the malolactic fermentation. After the ending of the malolactic fermentation, the wines are racked and submitted to ambient with low temperatures (-2 °C to 3 °C) for 10 days to promote the tartrate stabilization. This procedure involves the complexation of the sodium and calcium bitartrate salts by the action of the cold, i.e., it is characterized as a physical phenomenon involving low temperature. After the tartrate stabilization, the wines are then racked for the third time and then bottled (Figure 1) [10, 12, 13, 18].

Lee et al., (2006) [8] studied the chemical and sensory profile of wines produced form two Korean *Vitis labrusca* grapes, Gerbong and Campbell Early, and compared them with a French Beaujolais Nouveau wine. The winemaking procedure consisted of destemming and crushing the grapes using 200 ppm of sulfur dioxide and 0.2 g of dry yeast per liter. They added sugar into the must until it reached 21 °Brix to achieve the expected alcohol content. The resulted wines were once racked and aged in oak barrels. After this, the wines were stored at 11 °C for 6 months and finally bottled.

Studies with different *Vitis labrusca* and hybrid grape cultivars were developed analyzing the influence of the traditional and alternative (grape pre-drying and submerged cap) winemaking procedures on the phenolic, volatile, and sensory profile of Bordô [7, 23], Isabel [23], BRS Carmem [7], BRS Violeta [23], BRS Cora [14], and BRS Rúbea [14] red wines. The traditional winemaking procedure applied in these works followed the technique mentioned in Figure 2. The grapes were harvested in their optimal maturity stage and then were destemmed and crushed allowing for the release of the grape juice. The must was fermented by the inoculation of 20 $g.hL^{-1}$ of active dry *Saccharomyces cerevisiae* using 80 $g.hL^{-1}$ of sulfur dioxide. The wine musts were chaptalized to 11% v/v ethanol. The maceration lasted 7 days, pumping twice a day, and after this time, the wines were dejuiced. After dejuicing, the wines were racked three times, following the malolactic fermentation after the first racking.

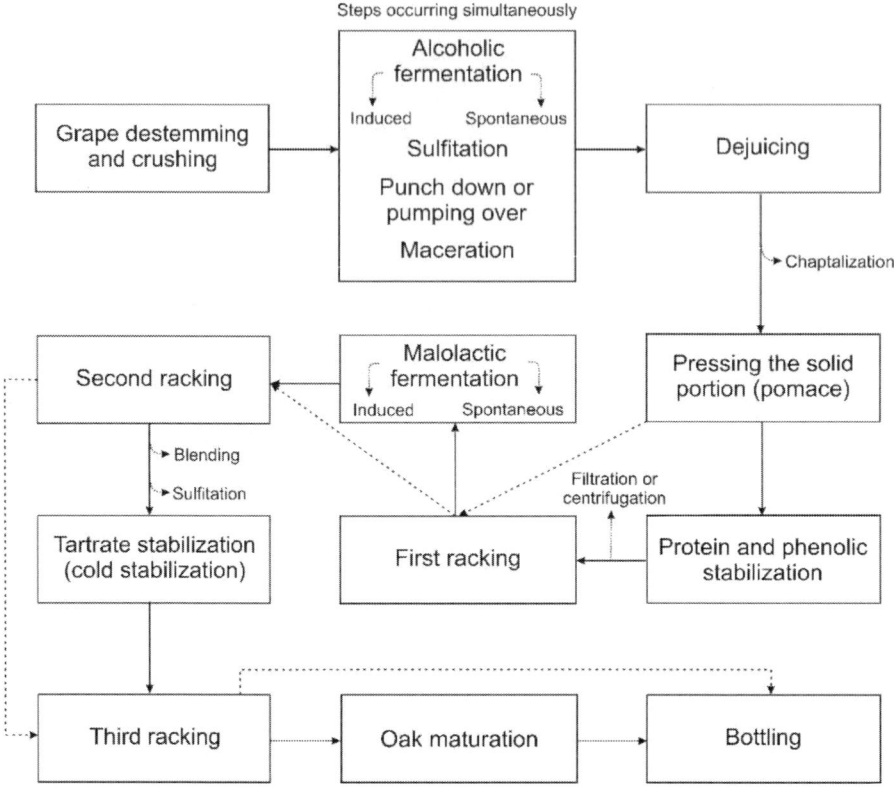

Figure 1. Traditional winemaking procedure. Full lines show the traditional winemaking method with all the steps, sectioned lines show the traditional winemaking procedure without some steps (Adapted from De Castilhos et al., (2016) [19]).

This second fermentation was induced by the inoculation of *Oenococcus oeni* lactic bacteria, and Thin Layer Chromatography (TLC) determined the final stage of the malolactic fermentation. The wines were then submitted to tartrate stabilization at a refrigerated ambient and then bottled. These works presented no filtration or finning process, and neither the use of pectolytic enzymes for the enhancement of the winemaking yield.

These works also analyzed two alternative winemaking techniques: grape pre-drying and submerged cap. The grape pre-drying consisted of drying the grapes before their destemming to enhance the solid soluble content avoiding the chaptalization process (Figure 3). This technique aimed at analyzing the influence of the grape drying on the wines' chemical and sensory profiles. After the grape pre-drying, the winemaking followed the traditional protocol.

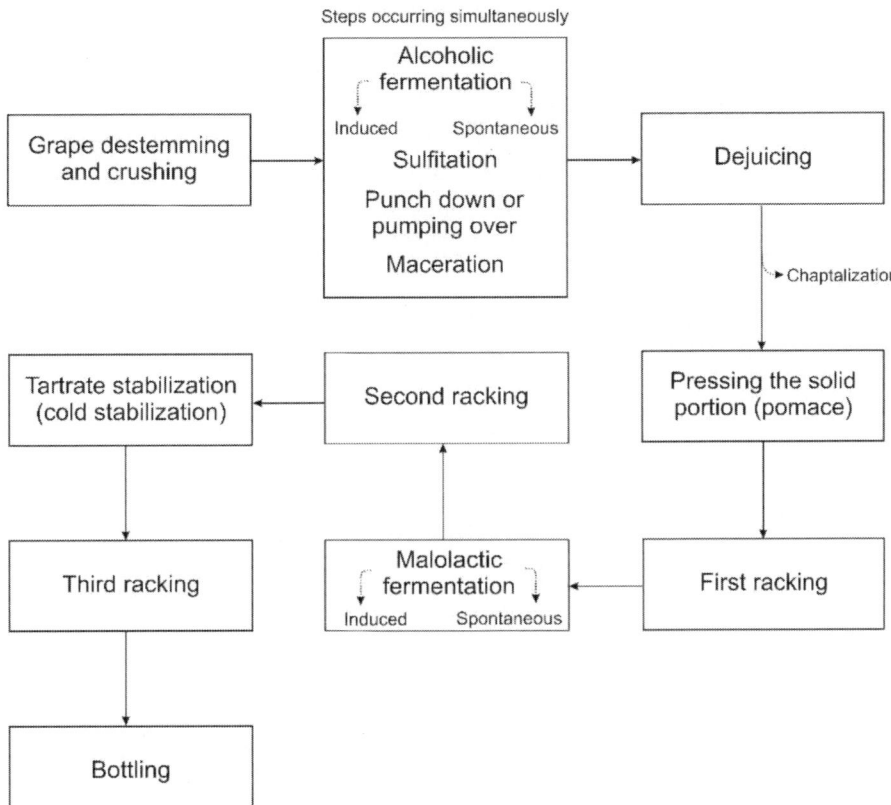

Figure 2. Traditional winemaking procedure (Adapted from De Castilhos & Del Bianchi (2016) [12]).

The other winemaking technique applied by the authors in all these mentioned works was the submerged cap. This winemaking procedure consisted of providing the constant maceration effect, i.e., promoting the constant contact between the must (liquid part) and the grape pomace (solid part) using stainless steel screens that have been adapted to the fermentation vessel (Figure 4). These screens avoided the rise of the grape pomace to the superior part of the fermentation vessel due to the formation of carbon dioxide, one of the biochemical products of the alcoholic fermentation. The submerged cap technique was developed during the maceration step. After the 7 days of maceration using the submerged cap technique, the wines followed the traditional winemaking protocol.

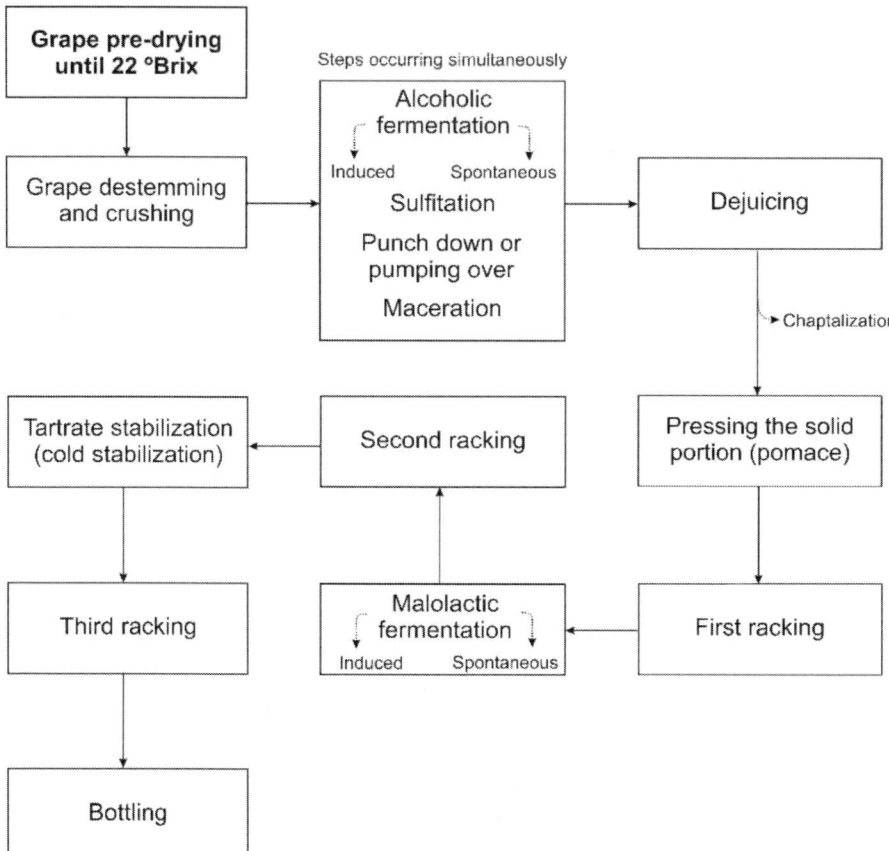

Figure 3. Grape pre-drying winemaking procedure (Adapted from De Castilhos & Del Bianchi (2016) [12]).

There is a lack of studies concerning the wines produced from *Vitis labrusca* grapes and hybrids since these grapes produce wines with fruity and sweet features, which are not very appreciated by the wine consumers and wine experts worldwide. However, these grapes present higher concentrations of phenolic compounds such as anthocyanins, flavonols, flavan-3-ols, and phenolic acids, enhancing their antioxidant properties. Also, the wines produced from American grapes present higher contents of higher alcohols and esters, volatile compounds that promote their fruity and sweet aroma and flavor.

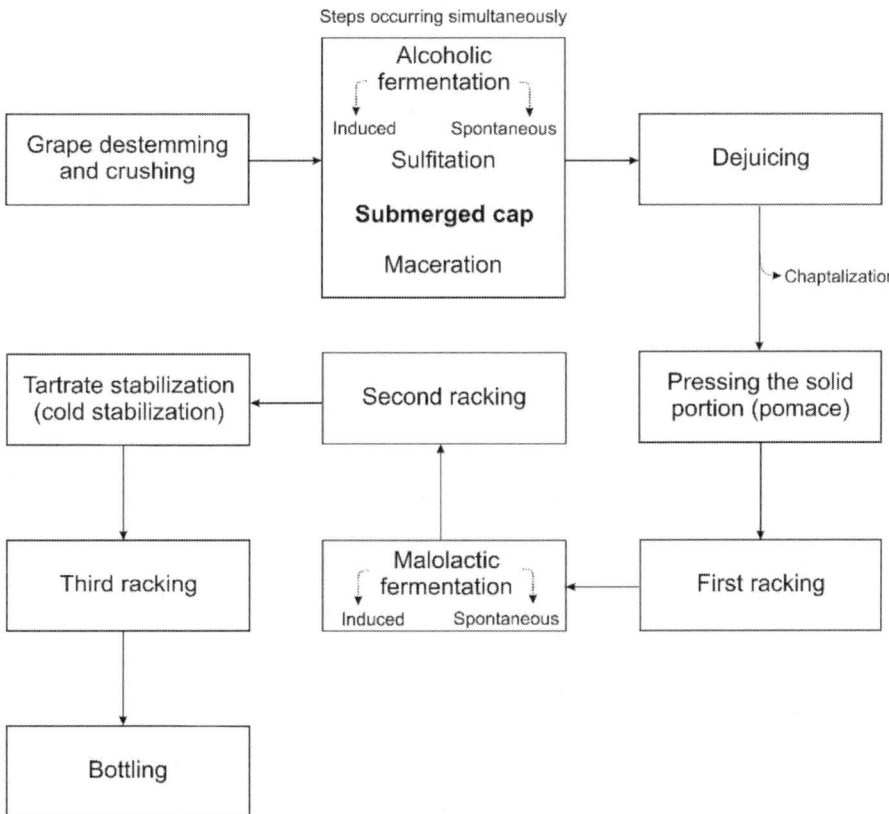

Figure 4. Submerged cap winemaking procedure (Adapted from De Castilhos & Del Bianchi (2016) [12]).

CHEMICAL AND SENSORY PROFILES

All the changes performed in the winemaking process will bring sensory modifications for wine. All the sensory wine attributes are closely related to a specific chemical compound or a group of chemical compounds. All the factors that influence the wine quality, such as vine management, soil, climate, water availability, sun exposure, nutrients of the soil, the winemaking procedure itself, will respond for a positive or a negative influence on the sensory wine profile. In this context, the variations in the winemaking process cause substantial changes in the wine chemical profile and, therefore, it causes significant changes in the sensory wine profile. Therefore, it is useful to analyze all the chemical changes that occurred in wine production and the resulted changes in their sensory profiles.

Lee et al., (2006) [8] analyzed the chemical and sensory profile of wines produced form two Korean *Vitis labrusca* grapes, Gerbong and Campbell Early, and compared them with French wine. The wines presented higher content of alcohols, highlighting the presence of isoamyl alcohol and 2-phenylethylethanol. The class of esters also presented high expression in those wines, accounting for a high concentration of ethyl hexanoate, ethyl decanoate, and 2-phenylethylacetate. These mentioned compounds are responsible for providing a fruity aroma and taste for the wines, a typical feature of the wines produced from *Vitis labrusca* grape cultivars [10, 20]. The authors also reported a higher concentration of volatile phenols, which are responsible for off-flavors in wine. The authors showed that these compounds were released from the oak aging since this group of chemical compounds is naturally found in oak barrels. However, volatile phenols also can be a result of biochemical degradation of grape phenolic acids by some specific yeasts, such as the Brettanomyces genus, and bacteria, accounting for possible wine contamination.

De Castilhos et al., (2020) [14] studied the volatile profile and sensory descriptive aroma profile of red wines produced from two *Vitis labrusca* grape cultivars: BRS Rúbea and BRS Cora. The authors also studied the application of two alternative winemaking procedures: grape pre-drying and submerged cap. The authors reported the higher concentration of ethyl and methyl esters (ethyl hexanoate and ethyl octanoate), dicarboxylic acid esters (diethyl succinate), terpenes (hydroxylinalool), and benzene compounds (2-phenyl ethanol), compounds that defined the wines as fruity and foxy. BRS Cora wines were defined as vegetal, and this sensory attribute was intensely correlated with C6 alcohols (1-hexanol and cis-3-hexenol). The wines produced from the pre-drying winemaking procedure presented no relationship with any sensory descriptive attribute; however, the presence of volatile furfural compounds in high concentrations accounted for an unpleasant aroma.

Racowski et al., (2018) [24] studied the effect of the application of different coatings on Niagara Rosada grapes to analyze their influence on wine production. The authors reported that the use of chitosan as a grape coating enhanced the phenolic concentration of the Niagara Rosada wines; in contrast, the grapes covered with Aloe vera coating presented the lowest concentration of phenolic compounds. Also, the authors reported the improvement of the shelf life of the grapes covered with chitosan coating, showing its potential as an alternative for the improvement of the grape shelf life without causing relevant changes on the grape and wine chemical properties.

Arcanjo (2015) [25] compared the color indexes, phenolic compounds, and antioxidant activity of different Isabel red wines (*Vitis labrusca*) according to their production region. The authors compared the wines produced in the South (traditional wine production region) with the Northeast region of Brazil. The wine produced in the Northeast region of Brasil presented two chemical parameters in disagreeing with the Brazilian legislation for wines: alcohol content and reducing sugars, showing a possible problem faced by the winery during the winemaking process. The Isabel red wines produced in the South of Brazil presented a higher concentration of anthocyanins, flavonols, and total phenolic compounds.

Biasoto et al., (2014) [20] analyzed the preference drivers of red wines produced from *Vitis labrusca* and hybrid grapes. The grapes that the authors studied were Ives (V. labrusca), Isabella (V. labrusca), Máximo (hybrid), Seibel 2 (hybrid), and Sanches (hybrid). The authors reported that the wines presenting the Ives grape in their composition showed more intense aroma/flavor notes described as sweet, grape, blackberry, and roses. In contrast, the wines produced from the Máximo grape were defined as earthy/mushroom, vegetative/green beans, woody, and yeast sensory notes. The sweet, grape-like, and blackberry notes were considered as preference drivers for the tasters; and the consumers did not like the earthy and vegetative sensory attributes for these wines. Consumers also dislike the sensory attributes that remember dried fruits as raisins. The authors also reported the highest phenolic concentration in the wines produced from the Máximo grape cultivar, corroborating with the "seed" flavor, astringency, bitterness, and body sensory attributes. The wines elaborated with Seibel 2, Ives, and Isabella were preferred by the majority of the consumers.

Burin et al., (2014) [26] evaluated the bioactive compounds and the antioxidant activity of *Vitis labrusca* grapes, Niágara, Niágara Rosada, Isabel, Concord, and Bordô, comparing them with Vitis vinifera grape cultivars using High-Performance Liquid Chromatography (HPLC). The authors reported that the polyphenols composition of the grapes is directly related to the different grape varieties instead of their species. They also reported a higher correlation between antioxidant activity and trans-resveratrol. The *Vitis labrusca* grape varieties presented high antioxidant capacity, associated with the presence of bioactive compounds, especially in Bordô and Concord grapes. The authors also reported that these grapes have great potential for winemaking to produce wines with high quality and relevant nutritional features.

De Castilhos et al., (2015) [7] studied the influence of alternative winemaking procedures in the phenolic and descriptive sensory profiles of red wines produced from Bordô and BRS Carmem grapes. The authors reported the formation of products resulted from the Maillard reaction that enhanced the antioxidant activity of the wines submitted to the grape pre-drying process. The wines produced from the pre-dried grapes also presented higher acidity, bitterness, and herbaceous taste. The wines produced from submerged cap winemaking presented an intense violet hue due to the presence of a high concentration of diglucoside, acetylated and coumaroylated anthocyanins. Also, the authors reported that the grape pre-drying caused an enhancement of the wine body.

In this same study, the authors used accurate instrumental techniques to identify and quantitate the phenolic wine compounds, such as High-Performance Liquid Chromatography coupled with Mass Spectrometry

(HPLC-MSn). The authors identified a possible new compound belonging to the anthocyanin class, known as malvidin 3-(6"-coumaroyl)-glucoside-5-(6"-coumaroyl)-glucoside since its spectra from Mass Spectrometry showed the loss of two 6"-coumaroyl-glucose moieties, suggesting that this loss occurred at C-3 and C-5 positions of malvidin anthocyanidin. Both the Bordô and BRS Carmem wines presented this compound in their composition. This result highlights the novelty of these grape cultivars in the enology field and the potential of these grapes in produce wines with high antioxidant capacity. The identification of this novel compound can be useful in the study of the antioxidant capacity and the color stability of the wines produced from these grapes.

CONCLUSION

The present chapter summarized the winemaking procedure that is applied in wines produced from *Vitis labrusca* grape cultivars to enhance the extraction of relevant compounds, primarily phenolic compounds, which are related to wine color, astringency, stability, and antioxidant capacity. The variation observed in the winemaking procedures was directly correlated to the changes in the chemical profile, which influence the sensory wine profiles. Also, the chemical and sensory profiles of wines produced from *Vitis labrusca* grapes showed that these wines have color potential, and they can be targeted to a unique public that consumes sweet and fruity wines due to the presence of a higher concentration of higher alcohols and esters. Studied showed that these wines have high antioxidant potential and stability, and, despite their unique (not too appreciated) sensory features, these wines deserve the world's attention due to their relevance for the consumers that seek an alcoholic beverage with compounds that causes positive effects for health. Considering that these grapes are cultivated in some punctual regions in the world, there is a lack of studies in this area; therefore, several studies still have to be developed to show the potential of these wines.

REFERENCES

[1] OIV. Organisation Internationale de la Vigne et du Vin. *World Viticulture situation*. 2019. [International Organization of Vine and Wine. *World Viticulture situation*]

[2] OIV. Organisation Internationale de la Vigne et du Vin. *World Viticulture situation*. 2018. [International Organization of Vine and Wine. *World Viticulture situation*]

[3] De Castilhos, M. B. M., Gómez-Alonso, S., García-Romero, E., Del Bianchi, V. L., & Hermosín-Gutiérrez, I. (2017). Isabel red wines produced from grape pre-drying and submerged cap winemaking: A phenolic and sensory approach. *LWT - Food Science and Technology*, *81*, 58–66.

[4] Oliveira, J. B., Egipto, R., Laureano, O., Castro, R., Pereira, G. E., & Ricardo-da-Silva, J. M. (2019). Chemical composition and sensory profile of Syrah wines from semiarid tropical Brazil – Rootstock and harvest season effects. *LWT - Food Science and Technology*, *114*, 108415.

[5] Panceri, C. P., De Gois, J. S., Borges, D. L. G., & Bordignon-Luiz, M. T. (2015). Effect of grape dehydration under controlled conditions on chemical composition and sensory characteristics of Cabernet Sauvignon and Merlot wines. *LWT - Food Science and Technology*, *63*, 228–235.

[6] Fandiño, M., Vilanova, M., Caldeira, I., Silvestre, J. M., Rey, B. J., Mirás-Avalos, J. M., & Cancela, J. J. (2020). Chemical composition and sensory properties of Albariño wine: Fertigation effects. *Food Research International*, *137*, 109533.

[7] De Castilhos, M. B. M., Corrêa, O. L. S., Zanus, M. C., Maia, J. D. G., Gómez-Alonso, S., García-Romero, E., Del Bianchi, V. L., & Hermosín-Gutiérrez, I. (2015). Pre-drying and submerged cap winemaking: effects on polyphenolic compounds and sensory descriptors. Part II: BRS Carmem and Bordô (*Vitis labrusca* L.). *Food Research International*, *76*, 697-708.

[8] Lee, S. J., Lee, J. E., Kim, H. W., Kim, S. S., & Koh, K. H. (2006). Development of Korean red wines using *Vitis labrusca* varieties: instrumental and sensory characterization. *Food Chemistry*, *94*, 385-393.

[9] De Castilhos, M. B. M., Del Bianchi, V. L., Gómez-Alonso, S., García-Romero, E., & Hermosín-Gutiérrez, I. (2019). Sensory descriptive and comprehensive GC–MS as suitable tools to characterize the effects of alternative winemaking procedures on wine aroma. Part I: BRS Carmem and BRS Violeta. *Food Chemistry*, *272*, 462–470.

[10] Jackson, R. S. *Wine science*: Principles and applications. San Diego: Academic Press, 5th edition, 2020.

[11] Change provisions of Law n. 7678 of November 8th, 1988 (2005).

[12] De Castilhos, M. B. M., & Del Bianchi, V. L. (2016). Winemaking Procedures and Their Influence on Wine Stabilization: Effect on the Chemical Profile. In *Recent Advances in Wine Stabilization and Conservation Technologies* (1º ed, p. 63–93). Nova Science Publisher.

[13] Ribéreau-Gayon, P., Glories, Y., Maujean, A., & Dubourdieu, D. *Handbook of Enology*: The Chemistry of Wine, Stabilization and Treatments. Chichester: John Wiley & Sons, 2nd edition, 2006.

[14] De Castilhos, M. B. M., Del Bianchi, V. L., Gómez-Alonso, S., García-Romero, E., & Hermosín-Gutiérrez, I. (2020). Sensory descriptive and comprehensive GC-MS as suitable tools to characterize the effects of alternative winemaking procedures on wine aroma. Part II: BRS Rúbea and BRS Cora. *Food Chemistry*, *311*, 126025.

[15] Steensels, J., Daenen, L., Malcorps, P., Derdelinckx, G., Verachtert, H., & Verstrepen, K. J. (2015). Brettanomyces yeasts—From spoilage organisms to valuable contributors to industrial fermentations. *International Journal of Food Microbiology*, *206*, 24–38.

[16] Smith, B. D., & Divol, B. (2016). Brettanomyces bruxellensis, a survivalist prepared for the wine apocalypse and other beverages. *Food Microbiology, 59*, 161–175.

[17] Crauwels, S., Steensels, J., Aerts, G., Willems, K. A., Verstrepen, K. J., & Lievens, B. (2015). Brettanomyces bruxellensis, essential contributor in spontaneous beer fermentations providing novel opportunities for the brewing industry. *Brewing Science, 68*, 110–121.

[18] De Castilhos, M. B. M., Conti-Silva, A. C., & Del Bianchi, V. L. (2012). Effect of grape pre-drying and static pomace contact on physicochemical properties and sensory acceptance of Brazilian (Bordo ˆ and Isabel) red wines. *European Food Research & Technology, 235*, 345–354.

[19] De Castilhos, M. B. M., Del Bianchi, V. L., & Hermosín-Gutiérrez, I. (2016). Influence of new trends in wine technology on the chemical and sensory profiles. In A. L. B. Pena, L. A. Nero, & S. D. Todorov (Orgs.), *Fermented Foods of Latin America: From traditional knowledge to innovative applications* (1º ed, p. 153–174). CRC Press Taylor & Francis Group.

[20] Biasoto, A. C. T., Netto, F. M., Marques, E. J. N., & Silva, M. A. A. P. (2014). Acceptability and preference drivers of red wines produced from *Vitis labrusca* and hybrid grapes. *Food Research International, 62*, 456–466.

[21] De Castilhos, M. B. M., Corrêa, O. L. S., Zanus, M. C., Maia, J. D. G., Gómez-Alonso, S., García-Romero, E., Del Bianchi, V. L., & Hermosín-Gutiérrez, I. (2015). Pre-drying and submerged cap winemaking: effects on polyphenolic compounds and sensory descriptors. Part I: BRS Rúbea and BRS Cora. *Food Research International, 75*, 374-384.

[22] De Castilhos, M. B. M., Garcia Maia, J. D., Gómez-Alonso, S., Del Bianchi, V. L., & Hermosín-Gutiérrez, I. (2016). Sensory acceptance drivers of pre-fermentation dehydration and submerged cap red wines produced from *Vitis labrusca* hybrid grapes. *LWT - Food Science and Technology, 69*, 82–90.

[23] Tavares, I. M. C., De Castilhos, M. B. M., Mauro, M. A., Ramos, A. M., Souza, R. T., Gómez-Alonso, S., Gomes, E., Da-Silva, R., Hermosín-Gutiérrez, I., & Lago-Vanzela, E. S. (2019). BRS Violeta (BRS Rúbea × IAC 1398-21) grape juice powder produced by foam mat drying. Part I: Effect of drying temperature on phenolic compounds and antioxidant activity. *Food Chemistry*, *298*, 124971.

[24] Racowski, I., Carlotti Filho, M. A. C., Mafra, P. C., & Massutti, J. (2018). Efeito da aplicação de revestimentos em uva Niágara (*Vitis labrusca*) na elaboração de vinho. *FTT Journal of Engineering and Business*, 54–66.

[25] Arcanjo, N. M. O. (2015). *Quality of red wine produced with grapes of the Isabel cultivar (Vitis labrusca) from two regions of Brazil (Northeast and South)* [Master Thesis]. Universidade Federal da Paraíba.

[26] Burin, V. M., Ferreira-Lima, N. E., Panceri, C. P., & Bordignon-Luiz, M. T. (2014). Bioactive compounds and antioxidant activity of Vitis vinifera and *Vitis labrusca* grapes: Evaluation of different extraction methods. *Microchemical Journal*, *114*, 155–163.

Chapter 2

PHENOLIC COMPOSITION OF BORDEAUX CABERNET SAUVIGNON AND MERLOT GRAPES AND WINES

Kleopatra Chira[1,2,3], Maria Reyes González-Centeno[1,2,3], Michael Jourdes[1,2] and Pierre-Louis Teissedre[1,2,]**

[1]University Bordeaux, ISVV, EA 4577, Œnologie,
Villenave d'Ornon, France
[2]INRAE, ISVV, USC 1366 Œnologie, Villenave d'Ornon, France
[3]Tonnellerie Nadalié, Ludon-Médoc, France

ABSTRACT

Bordeaux remains the most crucial appellation in France, in addition to the most prestigious places in the wine scenery owing to its reputation, to its financial and marketing organization. The last ten harvests confirmed that red wines account for 90% of the total production in the region with Cabernet Sauvignon and Merlot considered as the predominant varieties. Their wine quality is directly affected by phenolic

---

* Corresponding Author's E-mail: pierre-louis.teissedre@u-bordeaux.fr.

composition. Bordeaux Cabernet Sauvignon grapes and wines have a richer phenolic profile than Merlot. More precisely, Bordeaux Cabernet Sauvignon presented 52%, 28%, and 15% more total anthocyanins, phenolic compounds, and total tannins, respectively, than Bordeaux Merlot wines. Bordeaux Cabernet Sauvignon wines showed intermediate concentrations of individual ellagitannins when compared with Italian and USA Cabernet Sauvignon wines matured in the same oak barrels during 12 months. Cabernet Sauvignon and Merlot sensory properties such as astringency sensation and bitterness taste are correlated with ellagitannin and tannin composition. Moreover, during aging in oak barrels, ellagitannins participate in the wine oxygen consumption rate.

Keywords: Cabernet Sauvignon, Merlot, polyphenols, sensory impact

INTRODUCTION

A vineyard that has no history probably has no future. The Bordeaux wine-producing region fulfills this prerequisite, as documents and archaeological finds attest to its vineyards aged more than 2000 years and their Roman origins. Particularly, Bordeaux historians, producers, and merchants date the birth of the great local vineyards back to the 12th century. The Bordeaux winegrowing region, which is renowned all over the world, covers a surface of 116 000 hectares, 113 000 of which are entitled to appellation status (82% of the area being dedicated to red wines and the rest to white). Bordeaux vintners produce over 3.6 million hectolitres of wine annually. While today, more than three-quarters of the appellation vineyards are planted with red grape varieties, in 1961, 62% of these areas were devoted to the white wines production.

Bordeaux vineyards lie on bedrock composed of tertiary formations, which date back to over 50 million years. 'The river', as the Gironde is known in Bordeaux, divides the vineyards into left- and right-bank. The left bank involves Médoc and Graves with the most known Médoc appellations Pauillac, Margaux, Saint Estèphe, and Saint Julien, whereas the right-bank area includes the Libournais, Bourg and Blaye with Pomerol and Saint Emilion appellations being the most known appellations. Each of these areas is characterized by several terroirs, where the plurality of soil profiles (gravel, sand, clay, and limestone), climatic conditions, cultivars, and human practices produce specific wine signatures. Red Bordeaux wines are mainly blended. It is well-known that left-bank wine blends are dominated by Cabernet Sauvignon, whereas the right-bank mainly features Merlot [1] (Table 1).

Table 1. Black grape varieties of the Bordeaux wine region

Varieties	Gironde	Médoc	Graves	Libourne region	Entre deux mers
Cabernet Sauvignon	29%	52%	38%	7%	37%
Merlot	52%	34%	48%	68%	37%
Cabernet Franc	15%	10%	13%	24%	21%
Malbec	3%	≤ 1%	≤ 1%	1%	5%
Petit Verdot	≤ 1%	4%	0%	0%	0%

Adapted from Seguin et al. 2004 [1].

Cabernet Sauvignon is a black wine grape variety from Bordeaux derived from a crossing between Cabernet Franc and Sauvignon Blanc [2]. Small berries in small cylindrical-conical-shaped clusters characterize this late-budding variety with a long maturity period. It is particularly susceptible to grapevine trunk diseases (esca, eutypa, excoriose) and powdery mildew [3]. It yields between 2 and 14 t/ha depending on vine vigor [2]. Vigorous, upright, and highly fruit-bearing, it produces deeply colored and tannic wines that are particularly aromatic with scents and characteristic flavors of their region of production and in particular of the winemaking and aging techniques used. Typical of Bordeaux, Cabernet Sauvignon is now the fourth most grown red grape in the world, accounting for some 150,000 ha of vineyards, 20,000 ha of which are located in Gironde. Cabernet Sauvignon is often blended with other grapes because it has high tannicity and deep color, but it is not particularly round or fat. The Bordeaux blend approach, which blends Cabernet Sauvignon with Merlot, Cabernet Franc, Malbec and/or Petit Verdot, has primarily been adopted by the New World winemakers. However, Cabernet Sauvignon can be successfully blended with other grape varieties. Cabernet

Sauvignon also has good aging potential; the vegetal flavors progressively fade, allowing more complex flavors to develop. Its vines are widely distributed across the world, more precisely Cabernet Sauvignon was the second most-planted vine variety in 2015. It is mainly grown in China, France, Chile, the United States, Australia, Spain, Argentina, Italy, and South Africa [4]. Cabernet Sauvignon has gained popularity because of its durability, but also because of the tannin levels.

Merlot, means "*Le petit merle noir*" in French, is a black wine grape variety from Bordeaux that was incorporated into wines from that region at the beginning of the 19th century. It is an early-budding and early-flowering variety with an average maturity period and a tendency to over-ripen in warm climates. Merlot has a trailing growth habit, normal vigor, and is prone to a range of problems related to coulure [5]. It is sensitive to downy and powdery mildew, botrytis, cicadellidae, and drought [2]. With small berries and clusters, Merlot has good fertility and is suited to close pruning [5]. In Bordeaux, Merlot yields between 6 and 11 t/ha depending on vine vigor [2]. Merlot grapes produce round, structured and deeply coloured wines, and can be blended with more tannic wines to give balance. Present in 37 countries, in 2015 Merlot covered 266,000 ha, or 3% of the total world area under vines [4]. Comparing to Cabernet Sauvignon, Merlot is a quicker maturing variety, produces dark colored wines that are at once alcoholic and relatively subtle, and aromatically close in style to Cabernet dominated efforts. They are less tannic, less robust, and less long-lived than blends composed of Cabernet Sauvignon and Cabernet Franc. In the Bordeaux winegrowing region, planting has greatly increased since the winter frosts of 1956 and Merlot occupies significant areas of Saint Emilion (50% minimum), and an even more substantial portion of Pomerol (sometimes as much as 75%) [1] (Table 1).

The first time that Merlot grapes were used to make wine was back in the 1700s when a winemaker in France named it as an ingredient in the Bordeaux wine blend he had made. After that, the grape became very popular for its ability to add sweetness to the wine when combined with the country's favorite Cabernet Sauvignon.

PHENOLIC COMPOSITION OF BORDEAUX CABERNET SAUVIGNON AND MERLOT GRAPES

Tannins belong to phenolic compounds, one of the most important family groups for red wine, as they directly affect wine quality. They are located in grape solid parts (skins, seeds, stalks), and are responsible for the stabilization of wines, both color, and sensory characteristics, due to their astringent and bitter properties [6]. Condensed tannins or proanthocyanidins are oligomers or polymers of polyhydroxyflavan. The monomeric flavanols differ in their hydroxylation pattern in ring A and B and in the stereochemistry of C_3 position (Figure 1). They generally consist of (+)-catechin (C), (-)-epicatechin (EC), (-)-epigallocatechin (EGC) and (-)-epicatechin-3-O-gallate (ECG) linked by C(4)-C(6) or C(4)-C(8) inter-flavanoid bonds (Figure 2). Under standard winemaking practices, skin phenols are more readily extracted; however, as maceration increases (as with red wines), the seeds play an increasingly important role. They significantly contribute to astringent and bitter properties and wine color stability.

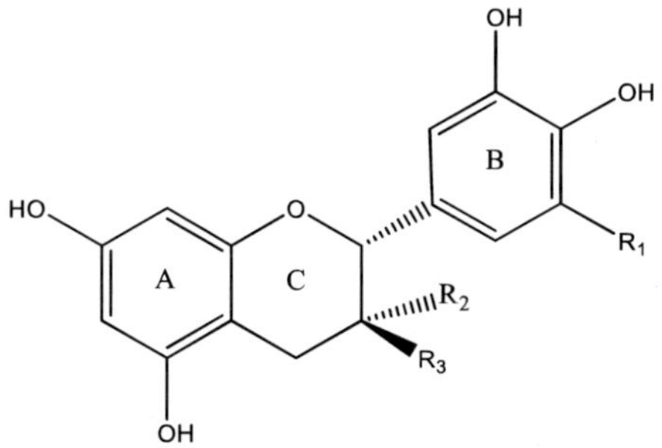

R_1=H R_2=H R_3=OH   (+)-Catechin
R_1=H R_2=OH R_3=H   (-)-Epicatechin
R_1=OH R_2=H R_3=OH   (+)-Gallocatechin
R_1=OH R_2=OH R_3=H   (-)-Epigallocatechin

Figure 1. Chemical structure of condensed tannins.

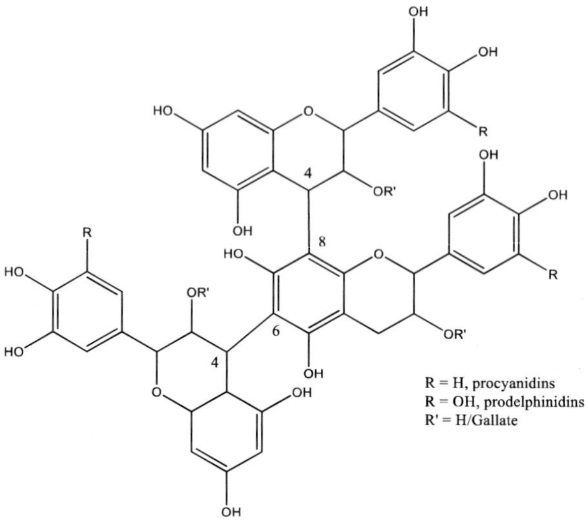

Figure 2. C(4)-C(6) or C(4)-C(8) inter-flavanoid bonds of proanthocyanidins.

The structure of tannins is characterized by the nature of its constitutive extension and terminal units, its mean degree of polymerization (mDP; the average number of units in the polymer), and its degree of galloylation (% G; the percentage of subunits bearing gallic acid esters). In the case of skins, for tannins characterization, the percentage of EGC, also called the percentage of prodelphinidins (% P), as well is used.

Cabernet Sauvignon has a tannic profile richer than Merlot. More precisely, in both seed and skin grape extracts a discrimination was achieved according to tannin composition with Cabernet Sauvignon being richer in catechin monomer having a greater mDP (mean degree of tannins polymerization) and % G (percentage of galloylation) than Merlot (Tables 2 and 3) [6-8]. Particularly, Cabernet Sauvignon seeds presented 22.2% and 6.5% higher mDP for 2009 and for 2006-2008 vintages respectively comparing to Merlot. However, the mean mDP values of Cabernet Sauvignon seeds were almost similar to those reported for the oligomeric fraction of Australian Cabernet Sauvignon seeds (mDP = 3.9). Greek Syrah seeds mDP (~ 2.0) was lower to Bordeaux Cabernet Sauvignon and closer to Merlot seeds [9, 10]. Regarding the percentage of galloylation for all the studied vintages, Cabernet Sauvignon presented almost 20% greater than

Merlot. As far as skin extracts are concerned, the same behavior was observed, Cabernet Sauvignon showed almost 20% higher mDP than Merlot. Even though Cabernet Sauvignon skin extracts from South Australia presented higher mDP levels (≥ 6.8), Greek Syrah skin oligomers showed lower mDP levels (≤ 2). Comparing to Bordeaux Merlot and Cabernet Sauvignon, the % P was almost 1.5 times more important in Cabernet Sauvignon than in Merlot. The dimers either in form of B1 or B2 or B3 or B4 (Figure 3) did not vary significantly according to the variety.

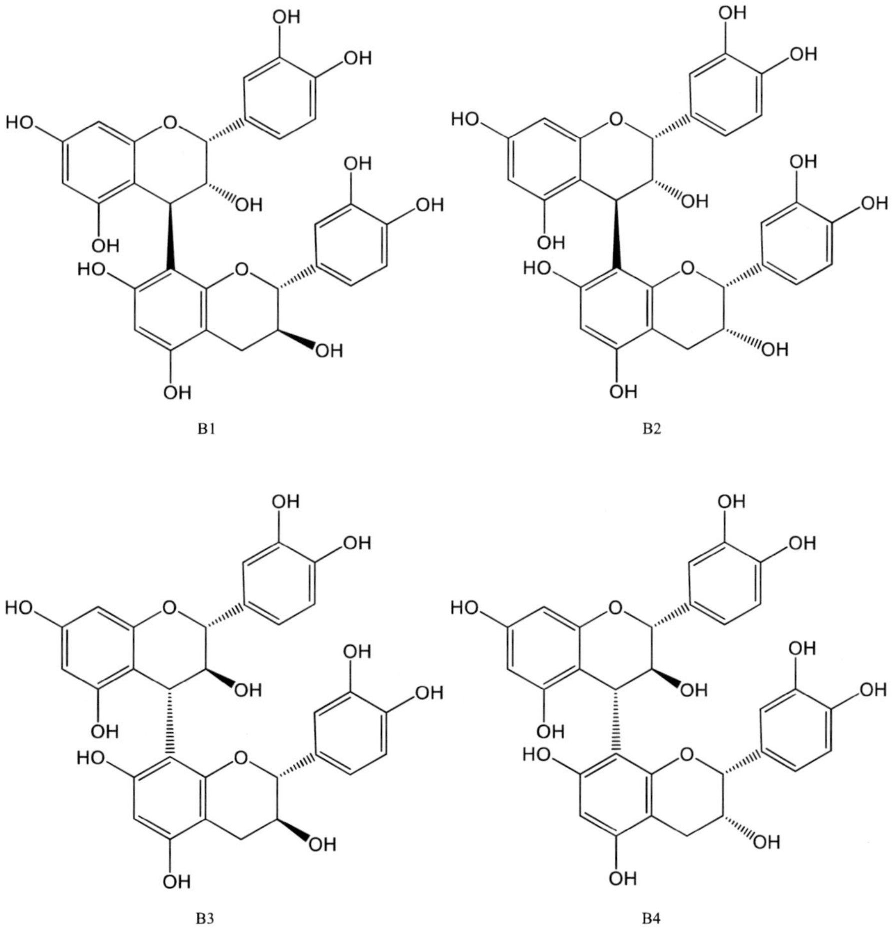

Figure 3. Chemical structure of procyanidins B1, B2, B3 and B4.

Table 2. Tannin characteristics of Bordeaux grape seed extracts in maturity [8, 12]

Grape seed characteristics		Grape variety			
		Cabernet Sauvignon[a]	Merlot[a]	Cabernet Sauvignon[b]	Merlot[b]
		2009	2009	2006-2008	2006-2008
	Total tannins	90.1 ± 4.0	92.2 ± 4.5	not reported	not reported
Monomers	C	1.73 ± 0.12	1.680 ± 0.117	3.81 ± 0.14	3.78 ± 0,16
	EC	1.35 ± 0.14	2.185 ± 0.151	2.07 ± 0.13	3.17 ± 0,08
	ECG (Epicatechin-3-O-gallate)	0.08 ± 0.005	0.271 ± 0.025	0.19 ± 0.04	0.24± 0.03
Dimers	B1	0.114 ± 0.006	0.100 ± 0.008	0.11 ± 0.02	0.43 ± 0.15
	B2	0.621 ± 0.051	0.530 ± 0.029	0.99 ± 0.11	0.80 ± 0.09
	B3	0.172 ± 0.011	0.233 ± 0.025	0.07 ± 0.02	0.07 ± 0.00
	B4	0.655 ± 0.053	0.295 ± 0.020	2.11 ± 0.16	2.44 ± 0.22
Trimer	C-C-EC	0.064 ± 0.001	0.331 ± 0.073	0.15 ± 0.02	0.27 ± 0.13
Mean Degree of Polymerization	mDP	2.7	2.1	3.3	3.1
Percentage of galloylation %	% G	28.3	23.6	24.5	20.0

[a] Lorrain et al. (2011) [8]; [b] Chira et al. (2009) [12]. C, Catechin; EC, Epicatechin; ECG, Epicatechin-3-O-gallate.

Condensed tannins with different mDPs may have different organoleptic properties. Generally, astringency increases with tannin concentration, molecular size, and % G [6]. Polymerized procyanidins are increasingly reactive with proteins and, therefore, have an increasingly astringent character [11]. Proanthocyanidins molecular size could also affect bitterness since monomers are considered bitterer than polymers. Therefore, the estimation of both mDP and % G of procyanidins could be a useful parameter to evaluate the type of procyanidins present in a sample.

Concerning the grape varieties and to the top growths of Graves and Médoc, sometimes it is impossible in blind tastings to differentiate the Cabernet Sauvignon and Merlot lots. In an effort to see if the sensory analysis of grape extracts could permit the differentiation of these two varieties, the results obtained are depicted in Table 4 [12]. Please leave one line space before Table 3

Table 3. Tannin characteristics of Bordeaux grape skin extracts in maturity [8, 12]

Grape skin characteristics		Cabernet Sauvignon[a]	Merlot[a]	Cabernet Sauvignon[b]	Merlot[b]
		2009	2009	2006-2008	2006-2008
	Total tannins	108.2 ±9.5	63.8 ±0.1	not reported	not reported
Monomers	C	0.168 ±0.005	0.047 ±0.000	0.31 ±0.06	0.19 ±0.06
	EC	0.061 ±0.001	0.030 ±0.000	0.25 ±0.13	0.18 ±0.07
	ECG	nd	nd	nd	nd
Dimers	B1	0.014 ±0.00	0.021 ±0.002	0.06 ±0.00	0.10 ±0,06
	B2	nd	nd	nd	nd
	B3	0.008 ±0.002	0.010 ±0.001	0.58 ±0.26	0.22 ±0,03
	B4	nd	nd	nd	nd
Trimer	C-C-EC	nd	nd	0.01 ±0.00	0.04 ±0.00
Mean Degree of Polymerization	mDP	5.2	4.3	22.9	19.4
Percentage of galloylation %	% G	11.8	27.7	3.24	1.3
Percentage of prodelphinidins %	% P	37.3	24.3	15.1	5.6

[a] Lorrain et al. (2011) [8]; [b] Chira et al. (2009) [12]. C, Catechin; EC, Epicatechin; ECG, Epicatechin-3-*O*-gallate; nd, not detected.

The panelist effect was not significant ($p > 0.05$), suggesting that homogeneous judges evaluated astringency and bitterness of tannin extracts. On a scale of intensity from 0-7 points, the average scores for astringency were 4.6 and 4.4 for seed tannin extracts, and 4.5 and 4.1 for skin tannin extracts in Cabernet Sauvignon and Merlot, respectively. For three successive vintages in Bordeaux, it was observed that the variety influenced neither the bitterness nor the astringency intensity.

A closer look at the data reveals a similar assessment of bitterness and astringency intensity, but in general, lower intensity values were attributed to Merlot astringency for both seed and skin extracts [12]. The average scores attributed by the judges for bitterness intensity were 3.8 and 4.2 in seed tannin extracts, and 3.1 and 3.6 in skin tannin extracts for Cabernet Sauvignon and Merlot, respectively. Nevertheless, this study concluded that vintage influenced both astringency and bitterness intensity significantly with extracts from 2016 being perceived as those with higher intensity of both sensory descriptors. It is worth pointing out that the concept of vintage is of the utmost importance in Bordeaux. There are many good, very good and even exceptional vintages, and there are many that are second-rate.

Except for tannins, another important phenolic compounds sub-family for the wine quality is anthocyanins. For the majority of red grape cultivars, such as Cabernet Sauvignon and Merlot, the anthocyanins accumulate mainly in the hypodermal cell layers of the berry skin after veraison. However, there are still a few select grapes of *Vitis vinifera* cultivars or their hybrids that can also accumulate high content of anthocyanins in their pulp. Anthocyanins play a key role in wine color intensity, hue, and stability. In *Vitis vinifera* grapes, anthocyanidins occur as 3-*O*-glucosides (anthocyanins), substituted with two (di-oxygenated: cyanidin- and peonidin-3-*O*-glucosides) or three (tri-oxygenated: delphinidin-, petunidin-, and malvidin-3-*O*-glucosides) hydroxyl (-OH) and/or methoxyl ($-OCH_3$) groups in the side-ring (B) of the flavonoid structure. Position 6 of glucose can also be esterified (acylated anthocyanins) by acetic (acetylated derivatives), *p*-coumaric (*p*-coumaroylated derivatives) and, less commonly, caffeic (caffeoylated derivatives) acids. Their structures are illustrated in Figure 4.

Table 4. Grape variety and vintage effect on astringency and bitterness intensity of seed and skin grape extracts [12]

Extract origin	Astringency intensity				
	Variety		Vintage		
	Cabernet Sauvignon	Merlot	2006	2007	2008
Seeds	4.6 ± 0.6 a	4.4 ± 0.2 a	5.0 ± 0.5 a	4.2 ± 0.1 b	4.4 ± 0.1 ab
Skins	4.4 ± 1.0 a	4.1 ± 0.3 a	5.3 ± 0.4 a	4.2 ± 0.1 b	3.3 ± 0.4 c
Extract origin	**Bitterness intensity**				
	Variety		Vintage		
	Cabernet Sauvignon	Merlot	2006	2007	2008
Seeds	3.8 ± 0.9 a	3.8 ± 0.3 a	4.2 ± 0.7 a	4.0 ± 0.3 a	3.3 ± 0.5 b
Skins	3.8 ± 1.2 a	3.8± 1.0 a	5.0 ± 0.3 a	3.3 ± 0.4 b	3.1 ± 0.3 b

a,b show the significant differences between varieties and among vintages (Duncan's test $p < 0.05$).

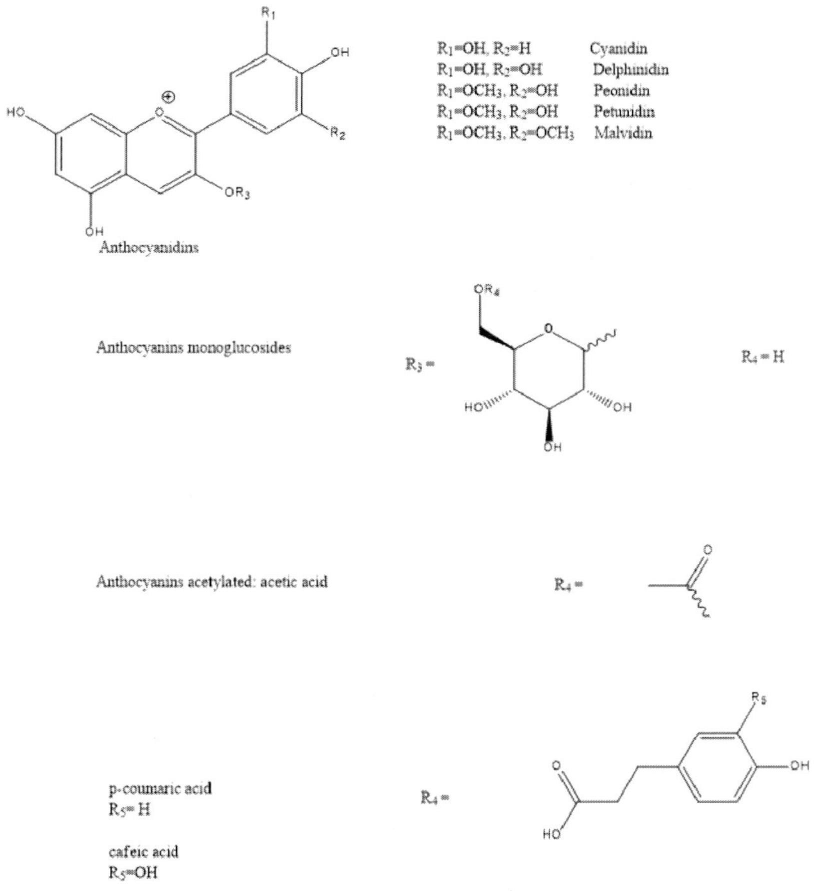

Figure 4. Chemical structure of anthocyanins.

The composition of anthocyanins (relative abundance of individual anthocyanins, the ratio of di-oxygenated vs. tri-oxygenated side-ring forms, the ratio of acylated vs. non-acylated derivatives, etc.) is variable among grapevine varieties [6, 13]. Indeed, acetylated derivatives levels are two times higher in Bordeaux Cabernet Sauvignon skin extract comparing to Merlot, whereas the coumaroylated derivatives of Merlot Bordeaux are 1.5 times greater than in Cabernet Sauvignon [6, 14, 15].

Regarding the malvidin, the predominant glycosylated anthocyanin, presented higher concentration in Bordeaux Cabernet Sauvignon, whereas peonidin was higher in Bordeaux Merlot (Table 5). The relation between the acetylated and coumaroylated forms allowed the differentiation of Cabernet Sauvignon grapes from Merlot in Bordeaux vineyards [6]. Comparing these two predominant Bordeaux varieties with Greek Cabernet Sauvignon and Merlot, the same tendency is observed, meaning that Cabernet Sauvignon is richer in malvidin than Merlot, whereas these two varieties along with Syrah and the two indigenous varieties Xinomavro and Agiorgitiko planted in Greece are in general richer in anthocyanins than Bordeaux varieties. However, looking at the anthocyanin composition of Portuguese planted autochthonous/non-autochthonous varieties, Bordeaux Merlot and Cabernet Sauvignon skins presented almost two or three times more individual anthocyanins. These significant variations are likely to be attributed to several factors that can affect anthocyanin biosynthesis, including sun exposure and other micro-climatic conditions, as well as fertilizing or maturation stage, among others.

Environmental and management conditions can significantly influence grape anthocyanin content; among these, irrigation is by far the most studied. A mild water deficit positively affects the accumulation of anthocyanins during ripening. It is attributed to multiple influences, either direct, such as enhanced expression of specific anthocyanin genes or indirect, such as reduced berry size and vegetative growth (the latter improving canopy microclimate and allocation of assimilates to berries). In general, anthocyanin concentration is maximized under non-irrigated conditions in all cultivars, but anthocyanin profile and relative distribution of individual anthocyanins among irrigation treatments are influenced principally by the cultivar [16].

Enzymatic activity that takes place in the anthocyanin biosynthesis is different in each variety [17]. Merlot seems to have a weaker flavonoid-hydroxylase activity, allowing the preferential accumulation of peonidin in the skins. Cabernet Sauvignon seems to have an intense activity in both flavonoid-hydroxylase and methyl-transferase activity, allowing the highest accumulation of malvidin. The profile and structure (e.g., esterification) of skin anthocyanins have an essential role in the anthocyanin interactions, such as self-association and co-pigmentation, thereby affecting the intensity and stability of red color in wine [18].

Table 5. Individual anthocyanins quantified in skin grape extracts

Bibliographic reference	Grape characteristics		Dp	Cy	Pt	Pn	Mv	Pn-acet+Mv-acet	Pn-coum+Mv-coum
	Grape variety	Geographical origin							
Chira et al. (2009)	Cabernet Sauvignon	Bordeaux	4.17 ± 0.26	0.84 ± 0.14	2.58 ± 0.14	3.24 ± 0.31	13.11 ± 1.07	6.99 ± 0.70	1.99 ± 0.16
	Merlot		4.55 ± 0.46	1.73 ± 0.21	3.62 ± 0.28	4.51 ± 0.49	10.83 ± 0.66	3.80 ± 0.35	3.31 ± 0.26
Kallithraka et al. (2009)	Cabernet Sauvignon*	Greece	9.96	1.44	10.57	11.78	106.88	not reported	14.81**
	Merlot*		9.37	2.95	11.74	14.61	79.91	not reported	18.34**
	Syrah*		6.26	1.48	10.48	16.01	103.45	not reported	48.23**
	Agiorgitiko*		3.97	0.97	5.98	11.28	97.64	not reported	39.66**
	Xinomavro*		0.55	0.17	1.47	3.06	23.86	not reported	9.59**
Costa et al. (2014)	Camarate	Dão/Portugal	nd	0.05 ± 0.01	0.22 ± 0.01	0.49 ± 0.02	5.08 ± 0.02	1.41 ± 0.08	1.01 ± 0.17
	Cabernet Franc	Douro/Portugal	0.95 ± 0.00	0.04 ± 0.00	1.03 ± 0.01	0.17 ± 0.02	4.10 ± 0.13	0.47 ± 0.01	1.54 ± 0.10
	Carignan Noir		0.34 ± 0.05	0.02 ± 0.01	0.57 ± 0.01	0.56 ± 0.03	5.60 ± 0.37	0.57 ± 0.02	3.79 ± 0.46
	Gamay		0.02 ± 0.01	nd	0.08 ± 0.02	0.12 ± 0.00	1.77 ± 0.03	0.27 ± 0.00	1.72 ± 0.03
	Tinta Amarela		0.58 ± 0.02	0.07 ± 0.01	0.74 ± 0.011	0.69 ± 0.02	2.91 ± 0.07	0.13 ± 0.01	0.71 ± 0.02

*Average values, **Mv-coum content.
nd: not detected; Dp, delphinidin-3-O-glucoside ; Cy, cyanidin-3-O-glucoside ; Pt, petunidin-3-O-glucoside, Pn, paeonidin-3-O-glucoside ; Mv, malvidin-3-O-glucoside ; Pn-acet, acetylated paeonidin-3-O-glucoside ; Mv-acet, acetylated malvidin-3-O-glucoside ; Pn-coum, coumaroylated paeonidin-3-O-glucoside ; Mv- coum, coumaroylated malvidin-3-O-glucoside.

The color intensity increases with the number of substituted groups on the B-ring (di-oxygenated forms are redder, while tri-oxygenated ones shift to blue) and with the replacement of hydroxyl by methoxyl groups (i.e., malvidin has the darkest color). Moreover, methoxylated anthocyanins (malvidin and peonidin) are more stable than hydroxylated ones to environmental and viticultural factors [19]. Thus, both Merlot and Cabernet Sauvignon skins dispose of the anthocyanins forms (i.e., malvidin and peonidin) that may better resist the degradation.

Both anthocyanin and tannin contents of wines depend on their concentrations in the grapes and on the vinification techniques used because maceration conditions determine the extraction of these compounds. Concurrently, during the vinification, some anthocyanins and tannins are adsorbed by the lees, precipitating with the tartaric salts. In contrast, other molecules undergo oxidation and hydrolysis, thus determining a significant decrease in the concentration of these compounds in the wine [20].

PHENOLIC COMPOSITION OF BORDEAUX CABERNET SAUVIGNON AND MERLOT WINES

Napoleon III and his entourage, among whom Baron Haussmann, were the power behind the conclusion of numerous free trade agreements which led different states and countries to almost totally suppress custom duties. The international trade in Bordeaux wine, thanks to the railway, could then be delivered in bulk to the North and the East of France and also to Paris. Until then, these regions had access only to wines from Burgundy, Argenteuil, and the Loire valley [21].

Once the cryptogamic disease of oidium had been mastered by the use of sulfur (1852-1856), the growing production of Bordeaux wines exceeded all previous limits. In terms of value, Bordeaux wine represented more than half of the French exports in 1863 and grew steadily between 1887 and 1889. This success was attributed to traditional clients from

Northern Europe (Scandinavia, Netherlands, Belgium, Germany, and especially Great Britain) as well as from North America and until 1890, from Argentina and Uruguay in South America. This latter area was populated by Mediterranean immigrants (Italian, Spanish, and even French origin), all of them being wine drinkers who did not have the time to develop their vineyards.

The best illustration of this prosperity lies in the vineyards' quality classification, such as the 1855 Classification. In 1855, the Chamber of Bordeaux Commerce (which included a large number of merchants) wanted to present the Gironde wines at the Universal Exhibition in Paris in the best possible circumstances. In order to avoid Bordeaux wines presented to be judged by persons with little or no knowledge, the Chamber of Commerce recommended that the wines should be left out of the competition and should be presented according to a specific classification. To this end, it was requested from the Syndicate of brokers a full list of the classified Gironde red wines, as well as one of the white *grands vins*. The classification was not based on a tasting process or a meticulous analysis, but on wines reputation, and their transaction prices: the wines with the highest prices have found themselves at the top of the list. This classification included only red wines from the Médoc, the Sauternes and Barsac sweet white wines, and one Graves red cru. This ranking was quite criticized because it has never been reviewed except in 1973 when the Château Mouton Rothschild was promoted from the rank of Deuxième Grand Cru Classé to that of Premier Grand Cru Classé (Médoc).

Despite their quality and their long-time reputation, Saint Emilion wines did not have their official classification until the end of the Second World War. It was in 1958, after many initiatives of the winegrowers of this region, which entitled the best growths to the specific "Grand Cru Classé" status. This classification was not based on the same criteria nor conducted under the same circumstances as the 1855 Classification. As the decree of 1958 provided that the Classification should be revised every ten years, it was updated in 1996, 2006, and most recently in late 2012. The Saint Emilion Classification is extremely strict and rigorous. The fact that it is revisable on a ten-year basis has led the producers to invest unrelentingly in their wine-entities in order to meet the required criteria as best as they can.

Except for Château Haut-Brion, the famous 1855 Bordeaux Classification did not consider any other Bordeaux wine from Graves appellation. To rectify that situation, the Graves appellation producers came up with their classification. The Graves classification took place almost 100 years after the official 1855 Classification of Bordeaux wine and emulated Médoc Classification in numerous ways. One of the key determining factors they took into consideration was the selling price over a long period as well as considerations for the terroir and soils in Pessac Léognan. The genesis for the 1959 Graves Classification came from the Graves Wine Syndicate. A select committee consisting of various wine merchants and brokers was brought together by the I.N.A.O, the *Institut National des Appellations d'Origine Controlée*. The members of the panel were asked to study the market price of the wine over a long period [22].

Owing to its reputation and both financial and marketing organization, Bordeaux remains the top appellation in France and the most prestigious in the world. The Château Mouton Rothschild in the heart of Médoc, Pauillac appellation, is one of the most popular in the region (Premier Grand Cru Classé) and produces the most prestigious wines having Cabernet Sauvignon as predominant variety. A comparative study based on chemical and sensory of Bordeaux Cabernet Sauvignon from Château Mouton Rothschild and of Bordeaux Merlot variety from Château la Providence (Pomerol) provided essential information for polyphenolic composition and sensory perception of Bordeaux predominant wine varieties. Wine chromatic parameters and characteristics like total phenolic, total tannin, total anthocyanins, for Cabernet Sauvignon, and Merlot are presented in Tables 6 and 7. Knowledge of the content and composition of phenolic compounds in monovarietal Bordeaux wines is of utmost importance and may be fundamental for adequate wine quality management and control. It can be useful to predict sensory wine properties and oxidative stability and to provide information on wine production technology and age.

For Cabernet Sauvignon wines from 24 vintages, total phenolic compounds ranged from 1579 to 3188 mg/L, whereas for Merlot wines ranked from 1244 to 2544 mg/L. Total tannins varied from 1.2-2.0 g/L and 1.2-2.1 g/L for Cabernet Sauvignon and Merlot, respectively. Meanwhile, total anthocyanin content fluctuated from 9.6 to 32.4 mg/L and from 15.3 to 12.4 mg/L for Cabernet Sauvignon and Merlot, respectively, for the most ancient to the most recent vintages. Bordeaux Cabernet Sauvignon presented 52%, 28%, and 15% more total anthocyanins, phenolic compounds, and total tannins, respectively than Merlot wines from Bordeaux. In young red wines, free anthocyanins are the principal source of red color, though monomeric anthocyanins are not particularly stable. As red grapes are the exclusive source of these monomeric anthocyanins, their composition can determine the anthocyanin profile of the corresponding red wines automatically and significantly [23].

These monomeric or free anthocyanins are gradually incorporated into their derived pigments, including copigments and polymeric pigments involving other phenolics during wine aging, contributing to a progressive shift of the red-purple color of young red wines towards the more red-orange color of aged red wines. It explains why color intensity for both varieties is decreased during aging, whereas the hue is increased (Tables 6 and 7). The hydroxylation pattern of the anthocyanins in the B ring can directly affect the hue and color stability due to the effect on the delocalized electrons path length in the molecule. For example, the anthocyanins with more hydroxyl groups in the B ring can contribute more blueness, whereas the degree of methylation of the B ring can increase the redness.

Thus, the malvidin-3-*O*-glucoside and its derivates are the reddest anthocyanins. Among these monomeric anthocyanins, malvidin-3-*O*-glucoside and its derivatives are usually the most abundant and are the source of most of the red color in very young red wines, varying from more than 90% in Grenache to just less than 50% in Sangiovese [24, 25]. In a recent study, malvidin-3-*O*-glucoside only accounted for no more than 42% of the total anthocyanins in a Merlot wine at the end of alcoholic fermentation [26]. In Bordeaux Cabernet Sauvignon for the late vintages, the malvidin-3-*O*-glucoside explained almost 83% of monomeric anthocyanins [6].

Table 6. Cabernet Sauvignon wine characteristics of different vintages [27]

Vintage	Phenolic compounds (mg AG/L)	Tannins (g/L)	Anthocyanins (mg/L)	CI	Hue
1978	1735.38 ± 15.11 ab	1.28 ± 0.00 ab	15.31 ± 0.62 a	0.46 ± 0.00 a	1.20 ± 0.02 k
1979	1806.6 ± 55.40 abc	1.65 ± 0.12 abc	33.25 ± 2.47 abc	0.45 ± 0.00a	1.14 ± 0.01ijk
1980	1778.12 ± 5.04 abc	1.62 ± 0.08 abc	31.06 ± 0.62 ab	0.44 ± 0.00a	1.08 ± 0.02 ghij
1981	1977.52 ± 5.04 abcde	1.30 ± 0.07 ab	29.31 ± 1.86 ab	0.44 ± 0.00a	1.19 ± 0.02 k
1982	2511.66 ± 20.14 fghi	2.14 ± 0.10 c	24.94 ± 0.62 ab	0.44 ± 0.00 a	1.08 ± 0.03 ghij
1983	2119.96 ± 35.25 bcdef	1.65 ± 0.15 abc	40.12 ± 3.53 abcde	0.44 ± 0.00 a	1.10 ± 0.04 ghijk
1984	1963.28 ± 15.11 abcd	1.72 ± 0.46 abc	25.81 ± 5.57 ab	0.43 ± 0.00 a	1.02 ± 0.00 fg
1985	1578.7 ± 35.25 a	1.17 ± 0.01 a	26.69 ± 0.62 ab	0.71 ± 0.00 b	1.07 ± 0.03 ghij
1987	2112.84 ± 10.07 abcdef	1.17 ± 0.12 a	27.13 ± 4.95 ab	0.94 ± 0.00 d	1.12 ± 0.02 ghij
1988	2411.96 ± 70.50 defgh	1.99 ± 0.05 c	46.38 ± 0.00 bcdef	1.11 ± 0.01 f	1.17 ± 0.02 jk
1992	2305.14 ± 5.04 cdefg	1.77 ± 0.21 abc	30.19 ± 1.86ab	1.18 ± 0.01 g	1.05 ± 0.01fghi
1993	2276.64 ± 45.32 cdefg	1.61 ± 0.23 abc	43.75 ± 3.71 bcdef	1.11 ± 0.02 f	1.02 ± 0.00 fgh
1994	2525.92 ± 0.00 fgh	1.92 ± 0.18bc	56.44 ± 1.86 cdef	1.03 ± 0.02 e	1.02 ± 0.04 fg
1995	2668.36 ± 30.22 ghi	2.07 ± 0.00 c	48.56 ± 4.33 bcdef	0.83 ± 0.02 c	1.03 ± 0.05 fgh
1996	2682.6 ± 50.36 ghi	2.11 ±0.33 c	45.94 ± 0.62 bcdef	1.03 ± 0.02 e	1.02 ± 0.04 fg
1997	2169.82 ± 80.57 bcdefgh	1.62 ± 0.11 abc	37.19 ± 9.28 abcd	1.25 ± 0.02 i	1.02 ± 0.03 fg
1998	2169.82 ± 80.57 defgh	1.94 ± 0.11 bc	48.13 ± 0.00 bcdef	1.26 ± 0.03 h	1.01 ± 0.03 fg
1999	2269.52 ± 120.86 bcdefg	2.09 ± 0.08 c	33.69 ± 6.81 abc	1.24 ± 0.02 h	0.95 ± 0.04 ef
2000	2917.62 ± 55.40 hi	2.06 ± 0.08 c	64.31 ± 3.09 dfg	1.29 ± 0.00 hi	0.89 ± 0.01 de
2001	2938.98 ± 110.79 hi	2.04 ± 0.08 c	61.69 ± 0.62 cdf	1.32 ± 0.01 ij	0.84 ± 0.02 cd
2002	3188.24 ± 25.18 i	2.13 ± 0.00 c	66.06 ± 0.62 fg	1.34 ± 0.00 j	0.83 ± 0.01 bcd
2003	2298.02 ± 40.29 cdefg	2.17 ± 0.23 c	87.06 ±15.47 g	1.35 ±0.00 jk	0.76 ± 0.01 bc
2004	2511.66 ± 120.86 efgh	2.10 ±0.07 c	116.38 ± 13.61 h	1.38 ± 0.00 k	0.73 ± 0.01 ab
2005	2903.38 ± 166.19 hi	2.23 ± 0.18 c	123.38 ± 14.85 h	1.48 ±0.00 l	0.65 ± 0.01 a

Mean values of two repetitions; a-k of each colon show the significant differences among vintages ($p \leq 0.05$); mg AG/L, mg/L of gallic acid; CI, color intensity.

Table 7. Merlot wine characteristics of different vintages [27]

Vintage	Phenolic compounds (mg AG/L)	Tannins (g/L)	Anthocyanins (mg/L)	CI	Hue
1979	1464.76 ± 35.3 ab	1.21 ± 0.01 a	9.63 ± 1.24 a	0.45 ± 0.00 a	1.16 ± 0.01 f
1986	1365.04 ± 85.61 a	1.31 ± 0.08 a	21.00 ± 2.47 ab	0.46 ± 0.00 a	1.11 ± 0.01 e
1988	1243.98 ± 20.1 a	1.32 ± 0.04 a	21.88 ± 1.24 ab	0.65 ± 0.00 b	1.02 ± 0.00 d
1994	2419.08 ± 126 cd	2.03 ± 0.05 c	80.94 ± 8.04 c	0.67 ± 0.00 c	0.98 ± 0.00 c
1995	1913.44 ± 20.1 bc	1.38 ± 0.1 ab	39.38 ± 12.37 b	0.70 ± 0.00 d	0.95 ± 0.00 b
1998	1714.02 ± 40.3 ab	1.58 ± 0.01 b	17.94 ± 9.28 ab	0.86 ± 0.00 e	0.66 ± 0.01 a
2003	2543.96 ± 42.3 d	2.09 ± 0.05 c	32.38 ± 2.47 ab	1.10 ± 0.00 f	0.68 ± 0.00 a

Mean values of two repetitions; a-f of each colon show the significant differences among vintages ($p \leq 0.05$); mg AG/L, mg/L of gallic acid; CI, color intensity.

Table 8. Comparison of tannin composition of Cabernet Sauvignon wines among different vintages [27]

Vintage	C (mg/L)	EC (mg/L)	B1 (mg/L)	B2 (mg/L)	B3 (mg/L)	B4 (mg/L)	T (mg/L)	mDP	% G	% P
1978	17.99 ± 0.54 abcde	5.71 ± 0.08 c	0.55 ± 0.01 ab	1.21 ± 0.08 ab	3.17 ± 0.03 bcd	1.38 ± 0.07 abcd	nd	1.81 ± 0.16 abc	1.37 ± 0.08 a	8.73 ± 0.38 abc
1979	12.46 ± 0.03 ab	3.41 ± 0.04 a	0.57 ± 0.01 abc	1.80 ± 0.1 ab	1.80 ± 0.0 a	0.98 ± 0.03 ab	Nnd	1.86 ± 0.04 a	2.99 ± 0.22 ab	16.09 ± 0.33 a
1980	16.03 ± 0.06 abc	4.02 ± 0.05ab	0.68 ± 0.00 bcde	3.13 ± 0.06 abc	2.94 ± 0.01 abc	1.73 ± 0.00 cde	nd	2.14 ± 0.09 abc	1.97 ± 0.34 cdef	12.00 ± 0.93 cdef
1981	7.82 ± 1.06 a	4.83 ± 0.01 bc	0.63 ± 0.00 abcd	1.14 ± 0.06 a	2.71 ± 0.01 ab	1.17 ± 0.04 abc	nd	1.85 ± 0.09 abcde	2.52 ± 0.22 bc	10.15 ± 2.05 abc
1982	17.02 ± 0.01abcd	5.54 ± 0.21 i	0.32 ± 0.02 a	9.60 ± 0.35 gh	2.47 ± 0.03 ab	0.99 ± 0.03 ab	nd	2.25 ± 0.14 ac	3.63 ± 0.12 cde	18.52 ± 0.98 ab

Table 8. (Continued)

Vintage	C (mg/L)	EC (mg/L)	B1 (mg/L)	B2 (mg/L)	B3 (mg/L)	B4 (mg/L)	T (mg/L)	mDP	% G	% P
1983	18.55 ± 0.88 abcde	7.17 ± 0.05 d	0.61 ± 0.10 abc	6.38 ± 0.14 ef	4.13 ± 0.58 cde	1.18 ± 0.09 abc	nd	2.40 ± 0.08 abcde	2.33 ± 0.05 fgh	17.62 ± 0.95 efgh
1984	23.41 ± 0.14 cdefgh	7.91 ± 0.01de	0.72 ± 0.03 bcde	2.61 ± 0.53 abc	4.55 ± 0.02 ef	0.87 ± 0.02 a	nd	2.52 ± 0.0 bcdef	4.63 ± 0.59 bcde	18.13 ± 0.08 defgh
1985	20.17 ± 1.41 bcde	7.18 ± 0.03 d	2.47 ± 0.07 k	7.50 ± 0.03 fg	4.48 ± 0.08 def	1.19 ± 0.07 abc	nd	2.02 ± 0.17 defg	3.26 ± 0.38 hij	14.64 ± 0.52 efgh
1987	25.72 ± 0.58 cdrgh	9.34 ± 0.17 f	0.95 ± 0.03 def	11.49 ± 0.01 hi	6.96 ± 0.11gh	1.51 ± 0.07 bcd	nd	2.86 ± 0.16 abcd	3.87 ± 0.47 dfgh	22.26 ± 0.23 bcde
1988	21.96 ± 0.29 bcdefg	8.00 ± 0.07 def	0.57± 0.01 abc	2.91 ± 0.13k	5.15 ± 0.12 ef	1.31 ± 0.02 abcd	nd	2.58 ± 0.09 fg	2.34 ± 0.27 fgh	12.82 ± 0.71 hi
1992	32.38 ± 0.54 gh	16.21± 0.53 i	1.36 ± 0.08 hi	16.1 ± 0.39 k	9.95 ± 0.17 lm	2.49 ± 0.29 gh	nd	2.37 ± 0.29 ef	3.30 ± 0.57 cd	24.28 ± 2.02 abcd
1993	20.96 ± 2.53 bd	9.16 ± 0.07 ef	0.71 ± 0.04 bcde	3.52 ± 0.05 bcd	5.21 ± 0.95 ef	1.24 ± 0.0 abcd	nd	3.68 ± 0.29 bde	4.91± 0.51ef	21.48 ± 2.99 ij
1994	26.93 ± 0.86 cdefgh	11.19 ± 1.35 g	1.12 ± 0.05 fg	5.81 ± 0.6 def	7.36 ± 0.05 h	2.14 ± 0.04 efg	nd	3.05 ± 0.06 i	4.22 ± 0.38 ij	20.72 ± 0.97 ghi
1995	31.49 ± 5.72 fgh	14.24 ± 0.09 h	1.32 ± 0.32 hi	16.60 ± 2.35 k	8.78 ± 0.83 jkl	3.34 ± 0.08 ik	nd	3.62 ± 0.11 gh	4.98 ± 0.45 ghi	28.01 ± 3.52 fghi
1996	34.00 ± 0.1 h	17.24 ± 0.01 ij	0.46 ± 0.03 ab	1.16 ± 0.0 a	9.10 ± 0.0 kl	1.31± 0.05 abcd	nd	3.68 ± 0.33 hi	5.80 ± 0.36 fgh	21.88 ± 2.47 defgh
1997	23.42 ± 0.01 cdefg	10.82 ± 0.05 f	0.88 ± 0.04 cdef	5.48 ± 0.4 def	5.66 ± 0.02fg	2.38 ± 0.05 fgh	nd	1.93 ± 0.14 j	0.88 ± 0.02 ij	11.16 ± 1.57 fgh
1998	22.83 ± 0.04 bcdefg	14.60 ± 0.06 h	1.64 ± 0.02 ij	17.85 ± 0.12 k	8.13 ± 0.6 hij	3.77 ± 0.08 k	0.10 ± 0.0	3.53 ± 0.12 ij	3.77 ± 0.1 dfg	17.30 ± 0.45 efg

Vintage	C (mg/L)	EC (mg/L)	B1 (mg/L)	B2 (mg/L)	B3 (mg/L)	B4 (mg/L)	T (mg/L)	mDP	% G	% P
1999	20.36 ± 0.02 bcde	8.40 ± 0.05 def	0.90 ± 0.02 cdef	4.36 ± 0.85 cde	5.22 ± 0.32 ef	2.15 ± 0.04 efg	nd	4.21 ± 0.20 ij	5.27 ± 0.33 dfgh	19.91 ± 1.51 defgh
2000	24.39 ± 1.18 cdefgh	11.61 ± 0.05 g	1.19 ± 0.15 fg	12.61 ± 0.2 i	7.61 ± 0.34 hi	1.82 ± 0.04 def	nd	3.94 ± 0.18 i	3.21 ± 0.2 ij	17.80 ± 0.74 j
2001	25.70 ± 0.12 cdefgh	14.29 ± 0.06 h	0.97 ± 0.01 ef	16.14 ± 0.04 k	7.33 ± 0.07 h	2.57 ± 0.13 gh	nd	3.99 ± 0.27 ij	3.20 ± 0.34 jk	17.22 ± 0.03 fghi
2002	31.90 ± 0.06 gh	24.60 ± 0.14 m	1.71 ± 0.05 j	22.11 ± 0.111	11.86 ± 0.06 n	5.02 ± 0.011	nd	7.62 ± 0.18 m	6.10 ± 0.1 k	18.66 ± 0.22 efgh
2003	26.79 ± 0.09 cdefgh	17.75 ± 0.08 j	0.70 ± 0.02 bcde	0.84 ± 0.02 a	8.98 ± 0.01 kl	2.76 ± 0.02 hi	nd	5.54 ± 0.1 k	3.41 ± 0.05 defgh	18.50 ± 0.01 efgh
2004	25.46 ± 0.35 cdefgh	20.55 ± 0.77 k	1.21 ± 0.08 fg	16.96 ± 0.02 k	11.27 ± 0.23 mn	3.49 ± 0.34 k	nd	6.63 ± 0.081	6.38 ± 0.33 k	21.61 ± 0.45 fghi
2005	28.38 ± 0.09 efgh	22.89 ± 0.061	1.17 ± 0.03 fg	15.71 ± 0.02 k	11.22 ± 0.01 mn	1.03 ± 0.08 ab	3.24 ± 0.41	7.13 ± 0.12 lm	5.7 ± 0.19 jk	18.46 ± 0.14 efgh

C, Catechin; EC, Epicatechin; nd: not detected.
Mean values of two repetitions; a-m of each colon show the significant differences among vintages ($p \leq 0.05$).

Table 9. Comparison of tannin composition of Merlot wines among different vintages (Chira et al. 2011)

Vintage	C(mg/L)	EC(mg/L)	B1(mg/L)	B2(mg/L)	B3(mg/L)	B4(mg/L)	T(mg/L)	mDP	% G	% P
1979	10.03 ± 2.09 b	3.01 ± 0.32 a	0.39 ± 0.16 ab	12.11 ± 8.87 a	0.91 ± 0.84 a	1.79 ± 0.72 a	nd	1.25 ± 0.03 e	1.42 ± 0.16 a	4.30 ± 0.72 b
1986	11.25 ± 0.46 b	4.12 ± 0.00 b	1.36 ± 0.04 d	13.88 ± 0.24a	0.60 ± 0.02 a	2.08 ± 0.02 a	nd	2.10 ± 0.06 a	1.53 ± 0.03 a	6.40 ± 1.10 b
1988	16.47 ± 0.54 c	6.81 ± 0.06 c	0.76 ± 0.19 a	16.36 ± 0.46 a	0.98 ± 0.14 a	1.25 ± 0.27 a	0.42 ± 0.07	2.45 ± 0.24 ab	2.23 ± 0.08 bc	15.10 ± 1.78 a
1994	31.39 ± 2.13 a	14.73 ± 0.17 d	0.69 ± 0.03a	9.53 ± 0.12 a	11.59 ± 0.57 c	0.91 ± 0.06 a	nd	2.64 ± 0.20 bc	1.80 ± 0.13 ab	17.50 ± 0.42 a
1995	21.34 ± 0.13 c	11.87 ± 0.02 e	0.09 ± 0.00 b	20.12 ± 0.04 a	3.04 ± 0.01 b	1.91 ± 0.03 a	nd	2.52 ± 0.20 ab	2.61± 0.32 c	16.10 ± 2.95 a
1998	30.06 ± 0.11 a	23.62 ± 0.09 f	6.81 ± 0.03 c	17.71 ± 0.03 a	0.27 ± 0.03 a	5.02 ± 0.03 b	nd	3.04 ± 0.19 cd	3.65 ± 0.08 d	19.70 ± 1.12 a
2003	31.19 ± 0.20 a	26.61 ± 0.56 g	6.97 ± 0.08 c	16.86 ± 1.23 a	1.00 ± 0.03a	8.38 ± 0.02 c	nd	3.43 ± 0.18 d	3.45 ± 0.19 d	18.40 ± 1.05 a

C, Catechin; EC, Epicatechin; nd: not detected.
Mean values of two repetitions; a-g of each colon show the significant differences among vintages ($p \leq 0.05$).

Tables 8 and 9 present the results concerning mDP, % G, % P, as well as tannin oligomer levels for Cabernet Sauvignon and Merlot wines. Among tannin monomers, (+)-catechin is the most predominant for both varieties. Among the dimers (B1, B2, B3, and B4), B2 and B3 are the principal ones with values of 22.1 mg/L and 11.9 mg/L respectively for Cabernet Sauvignon wines. For Merlot wines, B2 attained the highest level with a maximum of 17.7 mg/L, whereas B3 arrived at 11.6 mg/L. The same observation has been done for other varietal wines like Tempranillo, Graciano [28], whereas Syrah from Brazil presented less catechin (8.6 mg/L) than epicatechin (12.6 mg/L) [29].

Tannin mDP (Tables 8 and 9) ranged from 1.25 (vintage 1979) to 3.43 (vintage 2003) and from 1.81 (vintage 1978) to 7.13 (vintage 2005) for Merlot and Cabernet Sauvignon, respectively. Thus like in the case of seed and skin extracts, Cabernet Sauvignon exhibited almost 1.5 times greater mDP than Merlot. Other varietal red wines like Cabernet Franc and Sangiovese have been characterized with a mDP 4.9 and 9.8, respectively, [30] or even with a mDP for Carmenere to be 7.4 to 13.6 for 2004 and 2006 vintages. In contrast, Tinta Miuda red wine presented a mDP of 22.1 [11], and for an indigenous Greek red variety (Agiorgitiko), lower mDP values were reported, ranging from 1.43 to 2.27 [31]. Thus, mDP has different values depending on the grape variety, the vintage, but also the winemaking technique used. The differences among 23 vintages (1978-2005) for Cabernet Sauvignon, but even among the seven vintages for Merlot (1979-2005), were so significant that permitted to distinguish wines according to vintage.

Active wine consumers expect wine from a particular vintage and age to possess unique characteristics that differentiate it from other wines of the other vintages. These statements underline that a relationship between proanthocyanidin composition, sensory perception, and age of red wine may be essential to characterize wine quality. Indeed mDP decreases with wine bottle aging for both Cabernet Sauvignon and Merlot wines from Bordeaux [27, 32]. The oldest Cabernet Sauvignon wines (1978, 1979, and 1981) have a mDP that stays around 1.80; the young ones (1983–2001) have a mDP that varies from 2.0 to 4.2, and the youngest (2002–2005) possess a mDP between 5.5 and 7.6 values. The wines having a mDP between 2.0 and 4.0 are characterized as mellow and slightly astringent. A rougher sensation (tannic) is perceived for wines with a mDP higher than 4.0. It suggests that absolute mDP values may be helpful in characterizing wines not only according to their astringency but also according to their aging status and vintage.

The vintage concept is fundamental in Bordeaux. Wines produced from declassified harvests are generally sold at meager prices. It happened four times between 1963 and 1969 in 1991 in Saint Emilion (because of the heterogeneity of the harvest, which was due to spring frost) and in 1992 (because of hefty rainfall). Recently the vintage 2017 was characterized as a challenging vintage with considerable variation because of the frost in late April that devastated nearly half of the potential grape production. For the Château Mouton Rothschild, each vintage has its unique label. To celebrate its first vintage bottled at the château in 1924, Baron Philippe de Rothschild asked the famous poster artist Jean Carlu to create the very first label for Mouton Rothschild. This idea, however, would not be repeated until 1945, when a great contemporary painter designed the label of the new vintage. In that year, Baron Philippe decided to integrate a "V" as a symbol for the "victory of peace," and the label was drawn by the young painter Philippe Jullian. Thus, the honored tradition was born. Each year the prestigious château has invited a famous artist to grace the label of its celebrated wine with a unique design. Artists have included Dali, Caesar, Miro, Chagall, Picasso, Warhol, Soulages, Bacon, Balthus. Through the masterpieces of these great artistic minds, Mouton Rothschild has gradually expanded, year by year, its exciting and unique collection.

In Bordeaux, the majority of red wines undergo barrel-aging. The duration of barrel-aging, as well as the number of new barrels can vary from vintage to vintage. It is challenging to establish a standard protocol for aging wine in barrels since many possibilities and variables condition it. During oak wood contact, red wine undergoes important modifications, as spontaneous clarification, slow and continuous diffusion of oxygen through the wood pores of the oak barrel, and the extraction of many substances from the oak wood (e.g., aromatic compounds and ellagitannins) which modulate its organoleptic quality and complexity such as aroma, structure, astringency, bitterness, and color persistence. The ellagitannins (hydrolyzable tannins) are among these substances. In oak heartwood, they may represent 10% of the dry weight and are responsible for the high durability of this wood. Monomers, vescalagin, and castalagin are most predominant, and they represent between 40% and 60% by weight of the ellagitannins (Figure 5). The wood also extracts six other C-glycosidic ellagitannins into the wine: lyxose/xylose-bearing monomers grandinin and roburin E, the dimers roburin A and D, and the lyxose/xylose-bearing dimers roburins B and C (Figure 5). A study that evaluated and compared wines of Cabernet Sauvignon variety from different countries matured in the same oak barrels during 12 months showed that Bordeaux wines presented intermediate concentrations of individual ellagitannins when compared with Italian and USA Cabernet Sauvignon (Table 10).

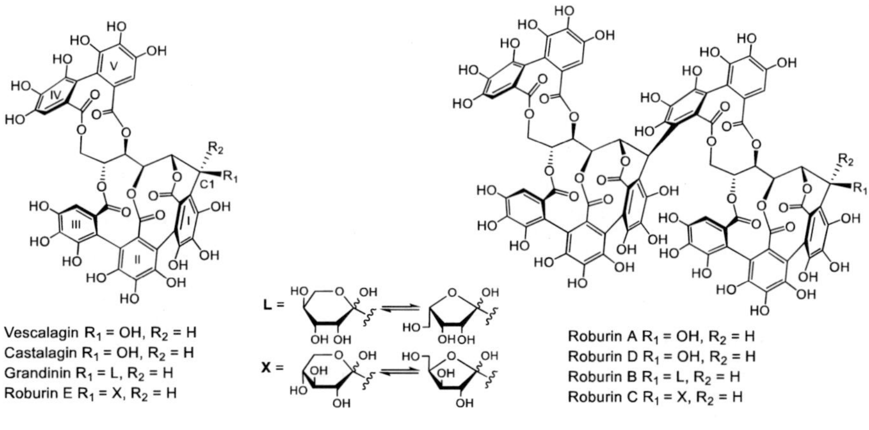

Vescalagin R_1 = OH, R_2 = H
Castalagin R_1 = OH, R_2 = H
Grandinin R_1 = L, R_2 = H
Roburin E R_1 = X, R_2 = H

Roburin A R_1 = OH, R_2 = H
Roburin D R_1 = OH, R_2 = H
Roburin B R_1 = L, R_2 = H
Roburin C R_1 = X, R_2 = H

Figure 5. Ellagitannins structure.

Regarding their sensory evaluation, French Cabernet Sauvignon wine being aged in medium toast barrels was characterized almost 1.5 times less astringent than both Italian and USA wines but with an intermediate bitterness on a 0 to 7 scale intensity [33] A direct implication of the ellagitannins concentration on the astringency and bitterness intensity ($p <$

5%) of a Bordeaux Cabernet Sauvignon wine being aged in oak barrels during 24 months was observed [34]. The same observation was made for a Bordeaux Merlot wine being in contact with derived oak wood products during one year [35].

Table 10. Ellagitannin composition of Cabernet Sauvignon wines from France, Italy and USA, aged 12 months in barrels according to González et al. [33]

Ellagitannin profile	Country		
	France	Italy	USA
Roburin A	0.08 ± 0.00	0.24 ± 0.01	0.12 ± 0.01
Roburin D	0.30 ± 0.01	0.91 ± 0.03	0.32 ± 0.04
Vescalagin	0.26 ± 0.01	0.70 ± 0.04	1.19 ± 0.09
Castalagin	3.43 ± 0.20	1.56 ± 0.03	8.38 ± 1.07
Roburins B+C	0.50 ± 0.03	1.56 ± 0.10	0.42 ± 0.02
Grandinin	0.24 ± 0.02	0.63 ± 0.08	0.16 ± 0.01
Roburin E	0.20 ± 0.02	0.17 ± 0.01	0.35 ± 0.03

Except for ellagitannins, oxygen is also considered as a critical aging parameter for wine quality. The moderate oxygen permeability of oak wood is probably the main reason why barrels are used for wine aging. Oxygen permeation through the wood causes different reactions between anthocyanins and proanthocyanidins, which stabilize the color and soften the astringency. Oxygen permeates the barrel through the pores in the wood and the joints between the staves [36], and it is assisted by the vacuum formed inside the barrels through the exchange of gases and the absorption of wine by the wood [37]. However, determining the exact oxygen transfer rate (OTR) of oak wood is not an easy matter, as is indicated by the widely varying OTR values. The most recent studies have used more reliable methodologies and point to an OTR of around 12 mg/L year for a 225 L barrel [36, 38].

Nevertheless, the oak wood oxygen consumption is not stable during wine aging; it has been reported that barrels OTR decreases over time once they have been filled [38]. For example, average oxygen concentrations in a Bordeaux red wine (95% Cabernet Sauvignon and 5% Cabernet Franc from Pessac Léognan region) revealed that oxygen consumption occurred in three stages. During the first eight days, once the barrels are filled, a rapid decrease in oxygen content was observed. Average oxygen concentrations in all the barrels decreased from 5138 ± 341 to 187 ± 25 μg dissolved oxygen/L red wine, with an average consumption rate of 618 ± 4 μg dissolved oxygen/L per day. In the second stage, oxygen consumption was slower, with levels decreasing from 187 ± 25 to 76 ± 4 μg dissolved oxygen/L in 97 days, at an average rate of 1.14 ± 0.38 μg dissolved oxygen/L per day. Then, when the winery temperature remained above 18°C, the wine's oxygen content was apparently below the limit of detection [39]. OTR is determined by wood moisture because oxygen diffuses better through dry wood [40]. Moreover, oxygen consumption was faster ($p < 0.05$) in Bordeaux wines, consisted of 95% of Cabernet Sauvignon, aged in barrels rich in ellagitannins comparing to wines aged in barrels paltry in ellagitannin levels. Suggesting a clear relationship between ellagitannin release and oxygen consumption rate [39].

CONCLUSION

Cabernet Sauvignon and Merlot (*Vitis vinifera L.*) are considered as ancient and traditional red wine grape varieties derived their fame from the southwest of France, Bordeaux. These varieties have become important red cultivars owing to their unique phenolic characteristics as well as to the research work relating to their quality and their wine production. Thus, the wines of the Bordeaux region benefit from the latest technical knowledge and new findings that, allied to a respect for the terroir, leads to constant improvement in their organoleptic quality. Even if the climate remains decisive in determine the quality and quantity of each vintage and even if accidents (damages caused by animals or plant parasites) occurring during the vine growing cycle, the scientific research as well as the application of new techniques and constant surveillance entertain the hope that Bordeaux will remain the most prestigious world appellation.

REFERENCES

[1] Seguin, G. (2004). The Terroirs of the Bordeaux Grand Crus. In C. Cocks (Ed.), *Bordeaux and its wines* (pp. 29-58). Bordeaux: Editions Feret.

[2] Galet, P. (2015). *Dictionnaire encyclopédique des cépages et de leurs synonymes*. (Libre & Solidaire ed.). [*Encyclopedic dictionary of grape varieties and their synonyms.*]

[3] Boursiquot, J. M., Charmont, S., Dufour, M. C., Moulliet, C., Ollat, N., Audeguin, L., Sereno, C., Desperrier, J. M., Jacquet, O., Lacombe, T., Leguay, M., & Schneider, C. (2007). *Catalogue des variétés et clones de vigne cultivés en France*. (IFV ed.). [*Catalog of vine varieties and clones cultivated in France.*]

[4] OIV. (2017). *Distribution of the world's grapevine varieties*. Paris.

[5] Galet, P. (1990). *Cépages et vignobles de France*. (Ministère de la Recherche et de la Technologie ed.). [*Grape varieties and vineyards of France*]

[6] Chira, K. (2009). *Structures moléculaires et perception tannique des raisins et des vins (Cabernet-Sauvignon, Merlot) du bordelais* (PhD Thesis). Université de Bordeaux 2, Bordeaux, France. [*Molecular structures and tannic perception of grapes and wines (Cabernet-Sauvignon, Merlot) from Bordeaux*]

[7] Chira, K., Lorrain, B., Ky, I., & Teissedre, P. L. (2011). Tannin composition of cabernet-sauvignon and merlot grapes from the bordeaux area for different vintages (2006 to 2009) and comparison to tannin profile of five 2009 vintage Mediterranean grapes varieties. *Molecules, 16*(2), 1519-1532.

[8] Lorrain, B., Chira, K., & Teissedre, P. L. (2011). Phenolic composition of Merlot and Cabernet-Sauvignon grapes from Bordeaux vineyard for the 2009-vintage: Comparison to 2006, 2007 and 2008 vintages. *Food Chemistry, 126*(4), 1991-1999.

[9] Kyraleou, M., Kallithraka, S., Theodorou, N., Teissedre, P.-L., Kotseridis, Y., & Koundouras, S. (2017). Changes in Tannin

Composition of Syrah Grape Skins and Seeds during Fruit Ripening under Contrasting Water Conditions. *Molecules, 22*(9), 1453.

[10] Teng, B., Hayasaka, Y., Smith, P. A., & Bindon, K. A. (2019). Effect of Grape Seed and Skin Tannin Molecular Mass and Composition on the Rate of Reaction with Anthocyanin and Subsequent Formation of Polymeric Pigments in the Presence of Acetaldehyde. *Journal of Agricultural and Food Chemistry, 67*(32), 8938-8949.

[11] Sun, B., Sá, M. d., Leandro, C., Caldeira, I., Duarte, F. L., & Spranger, I. (2013). Reactivity of Polymeric Proanthocyanidins toward Salivary Proteins and Their Contribution to Young Red Wine Astringency. *Journal of Agricultural and Food Chemistry, 61*(4), 939-946.

[12] Chira, K., Schmauch, G., Saucier, C., Fabre, S., & Teissedre, P. L. (2009). Grape variety effect on proanthocyanidin composition and sensory perception of skin and seed tannin extracts from Bordeaux wine grapes (cabernet Sauvignon and merlot) for two consecutive vintages (2006 and 2007). *Journal of Agricultural and Food Chemistry, 57*(2), 545-553.

[13] Kallithraka, S., Mohdaly, A. A. A., Makris, D. P., & Kefalas, P. (2005). Determination of major anthocyanin pigments in Hellenic native grape varieties (*Vitis vinifera* sp.): Association with antiradical activity. *Journal of Food Composition and Analysis, 18*(5), 375-386.

[14] Costa, E., Cosme, F., Jordão, A. M., & Mendes-Faia, A. (2014). Anthocyanin profile and antioxidant activity from 24 grape varieties cultivated in two Portuguese wine regions. *Oeno One, 48*(1), 51-62.

[15] Kallithraka, S., Aliaj, L., Makris, D. P., & Kefalas, P. (2009). Anthocyanin profiles of major red grape (*Vitis vinifera* L.) varieties cultivated in Greece and their relationship with in vitro antioxidant characteristics. *International Journal of Food Science & Technology, 44*(12), 2385-2393.

[16] Theodorou, N., Nikolaou, N., Zioziou, E., Kyraleou, M., Kallithraka, S., Kotseridis, Y., & Koundouras, S. (2019). Anthocyanin content and composition in four red wine grape cultivars (*Vitis vinifera L.*) under variable irrigation. *Oeno One, 53*(1), 39-51.

[17] Roggero, J. P., Coen, S., & Ragonnet, B. (1986). High Performance Liquid Chromatography survey on changes in pigment content in ripening grapes of Syrah. An approach to anthocyanin metabolism. *American Journal of Enology and Viticulture, 37*, 77-83.

[18] Han F. L., & Xu Y. (2015). Effect of the structure of seven anthocyanins on self-association and colour in an aqueous alcohol solution. *South African Journal of Enology and Viticulture, 36*(1), 105-116.

[19] Castellarin, S. D., & Di Gaspero, G. (2007). Transcriptional control of anthocyanin biosynthetic genes in extreme phenotypes for berry pigmentation of naturally occurring grapevines. *BMC Plant Biology, 7*(46).

[20] Rodrigues, A., Ricardo-Da-Silva, J. M., Lucas, C., & Laureano, O. (2013). Effect of winery yeast lees on Touriga Nacional red wine color and tannin evolution. *American Journal of Enology and Viticulture, 64*(1), 98-109.

[21] Roudie, P. (2004). The history of Bordeaux vineyards. In *Bordeaux and its wines* (pp. 17-28). Bordeaux: Edition Feret.

[22] Boidron, B. (2004). The classifications. In C. Cocks (Ed.), *Bordeaux and its wines* (pp. 165-171). Bordeaux: Editions Feret.

[23] Chira, K., Pacella, N., Jourdes, M., & Teissedre, P. L. (2011). Chemical and sensory evaluation of Bordeaux wines (Cabernet-Sauvignon and Merlot) and correlation with wine age. *Food Chemistry, 126*(4), 1971-1977.

[24] Gonzalez-Neves, G., Franco, J., Barreiro, L., Gil, G., Moutounet, M., & Carbonneau, A. (2007). Varietal differentiation of Tannat, Cabernet-Sauvignon and Merlot grapes and wines according to their anthocyanic composition. *European Food Research and Technology, 225*(1), 111-117.

[25] Castillo-Muñoz, N., Fernández-González, M., Gómez-Alonso, S., García-Romero, E., & Hermosín-Gutiérrez, I. (2009). Red-Color Related Phenolic Composition of Garnacha Tintorera (*Vitis vinifera* L.) Grapes and Red Wines. *Journal of Agricultural and Food Chemistry, 57*(17), 7883-7891.

[26] He, F., He, J. J., Pan, Q. H., & Duan, C. Q. (2010). Mass-spectrometry evidence confirming the presence of pelargonidin-3-O-glucoside in the berry skins of Cabernet Sauvignon and Pinot Noir (*Vitis vinifera L.*). *Australian Journal of Grape and Wine Research, 16*(3), 464-468.

[27] Puértolas, E., Álvarez, I., & Raso, J. (2011). Changes in phenolic compounds of Aragón red wines during alcoholic fermentation. *Food Science and Technology International, 17*(2), 77-86.

[28] Gómez-Alonso, S., García-Romero, E., & Hermosín-Gutiérrez, I. (2007). HPLC analysis of diverse grape and wine phenolics using direct injection and multidetection by DAD and fluorescence. *Journal of Food Composition and Analysis, 20*(7), 618-626.

[29] de Oliveira, J. B., Egipto, R., Laureano, O., de Castro, R., Pereira, G. E., & Ricardo-da-Silva, J. M. (2019). Chemical composition and sensory profile of Syrah wines from semiarid tropical Brazil – Rootstock and harvest season effects. *LWT - Food Science & Technology, 114*.

[30] Gris, E. F., Mattivi, F., Ferreira, E. A., Vrhovsek, U., Pedrosa, R. C., & Bordignon-Luiz, M. T. (2011). Proanthocyanidin profile and antioxidant capacity of Brazilian *Vitis vinifera* red wines. *Food Chemistry, 126*(1), 213-220.

[31] Petropoulos, S., Kanellopoulou, A., Paraskevopoulos, I., Kotseridis, Y., & Kallithraka, S. (2017). Characterization of grape and wine proanthocyanidins of Agiorgitiko (*Vitis vinifera* L. cv.) cultivar grown in different regions of Nemea. *Journal of Food Composition and Analysis, 63*, 98-110.

[32] Chira, K., Jourdes, M., & Teissedre, P. L. (2012). Cabernet sauvignon red wine astringency quality control by tannin characterization and polymerization during storage. *European Food Research and Technology, 234*(2), 253-261.

[33] González-Centeno, M. R., Chira, K., & Teissedre, P. L. (2016). Ellagitannin content, volatile composition and sensory profile of wines from different countries matured in oak barrels subjected to different toasting methods. *Food Chemistry, 210*, 500-511.

[34] Michel, J., Jourdes, M., Le Floch, A., Giordanengo, T., Mourey, N., & Teissedre, P. L. (2013). Influence of wood barrels classified by NIRS on the ellagitannin content/composition and on the organoleptic properties of wine. *Journal of Agricultural and Food Chemistry, 61*(46), 11109-11118.

[35] Chira, K., & Teissedre, P. L. (2013). Extraction of oak volatiles and ellagitannins compounds and sensory profile of wine aged with French wine woods subjected to different toasting methods: Behaviour during storage. *Food Chemistry, 140*(1-2), 168-177.

[36] Del Alamo-Sanza, M., Cárcel, L. M., & Nevares, I. (2017). Characterization of the oxygen transmission rate of oak wood species used in cooperage. *Journal of Agricultural and Food Chemistry, 65*(3), 648-655.

[37] Moutounet, M., Mazauric, J. P., Saint-Pierre, B., & Hanocq, J. F. (1998). Gaseous exchange in wines stored in barrels. *Journal des Sciences et Techniques de la Tonnelerie, 4*, 131-145.

[38] Nevares, I., & Del Alamo-Sanza, M. (2015). Oak stave oxygen permeation: A new tool to make barrels with different wine oxygenation potentials. *Journal of Agricultural and Food Chemistry, 63*(4), 1268-1275.

[39] Michel, J., Albertin, W., Jourdes, M., Le Floch, A., Giordanengo, T., Mourey, N., & Teissedre, P. L. (2016). Variations in oxygen and ellagitannins, and organoleptic properties of red wine aged in French oak barrels classified by a near infrared system. *Food Chemistry, 204*, 381-390.

[40] del Alamo-Sanza, M., & Nevares, I. (2014). Recent advances in the evaluation of the oxygen transfer rate in oak barrels. *Journal of Agriculture and Food Chemistry, 62*, 8892-8899.

In: Fermented and Distilled
Editors: M. B. M. de Castilhos et al. © 2021 Nova Science Publishers, Inc.
ISBN: 978-1-53618-985-8

Chapter 3

MALOLACTIC FERMENTATION OF TEMPRANILLO WINES: EFFECTS ON CHEMICAL COMPOSITION AND SENSORY QUALITY

Pedro Miguel Izquierdo-Cañas[1,2,*],
Sergio Gómez-Alonso[3], *Esteban García-Romero*[1],
Gustavo Cordero-Bueso[4]
and María de los Llanos Palop-Herreros[5]

[1]Instituto Regional de Investigación y Desarrollo Agroalimentario y Forestal de Castilla-La Mancha (IRIAF), IVICAM, Tomelloso, Spain
[2]Fundación Parque Científico y Tecnológico de Castilla-La Mancha, Albacete, Spain
[3]Instituto Regional de Investigación Científica Aplicada, Universidad de Castilla-La Mancha, Ciudad Real, Spain
[4]Departamento de Biomedicina, Biotecnología y Salud Pública (área de Microbiología), Universidad de Cádiz, Cádiz, Spain

---

[*] Corresponding Author's E-mail: pmizquierdo@jccm.es.

[5]Facultad de Ciencias Ambientales y Bioquímica.
Universidad de Castilla-La Mancha, Toledo, Spain

Abstract

This chapter describes changes that occurred in the enological parameters, volatile fraction composition, and sensorial quality of Tempranillo wines as a result of malolactic fermentation (MLF). The first part is dedicated to evaluating the influence of the strain of lactic acid bacteria (LAB) inoculated to perform the MLF for what three *Oenococcus (O.) oeni* strains were assayed: an autochthonous strain, C22L9, isolated from a winery in Castilla-La Mancha (Spain), and two other commercial strains, PN4™ and Alpha™ (Lallemand Inc.), all introduced by direct inoculation (MBR™). Strain C22L9 carried out MLF slightly faster than the two other commercial strains, leading to a lower increase in volatile acidity and 2,3-butanedione and 3-hydroxy-2-butanone concentrations, higher lactic acid content, and lower degradation of citric acid. The second part of the chapter is dedicated to evaluating the pros and cons of co-inoculation (COI) of LAB and yeast versus the traditional process carried out in wineries in which LAB are inoculated after completion of alcoholic fermentation by yeast, in a process known as sequential inoculation (SEQ). The study was performed over two commercial yeast strains (VRB™ and VN™) and an autochthonous *Oenococcus oeni* strain (C22L9), and parameters analyzed include the kinetic of vinification process and the chemical and sensory characteristics of Tempranillo wines produced. Results from this research showed that concurrent yeast/bacteria inoculation of musts produced a significant reduction of process length, without a pronounced degradation of L-malic acid during AF, nor an excessive increase in volatile acidity.

Keywords: tempranillo, malolactic fermentation, volatile composition, sensory profile

Introduction to Malolactic Fermentation (MLF)

Winemaking is a complex process that frequently involves two successive fermentations; an alcoholic fermentation (AF), conducted by yeasts, and a subsequent malolactic fermentation (MLF) carried out by wine lactic acid bacteria (LAB). MLF is generally considered to be a desirable transformation in the winemaking process, particularly in red wine production, in which L-malic acid is decarboxylated into L-lactic acid and CO_2, yielding de-acidification and more excellent microbiological stability of the wine. Besides, many other secondary metabolic reactions occur, producing changes in the sensory wine profile, which have significant consequences for their final quality [1-3]. MLF may occur spontaneously from the activity of native LAB naturally present in grapes and cellars, but sometimes the process takes weeks, and it does not always achieve satisfactory results [4] . Previous reports have shown the presence of different species and strains of LAB in spontaneous MLF, although *Oenococcus (O.) oeni* has been described as the predominant species [1, 5].

An oenological practice used as an alternative to the spontaneous MLF in some wineries is the inoculation of commercial malolactic starter cultures, which allows better control of how and when this fermentation takes place [6]. However, the use of starter cultures is not always successful because wine is a very harsh environment for bacterial growth [7]. Different environmental factors, such as the physicochemical parameters, the presence of energy sources, and the existence of other microbiota in the wine [8], can influence the implantation and growth of the inoculated bacteria and, therefore, the time required to complete MLF.

The use of MLF starter cultures comprising LAB strains selected from the indigenous wine microbiota of each region takes advantage of the natural adaptation of the strains to the wine characteristics, and may simultaneously preserve the characteristics of regional wines [9]. Strict criteria are used for the selection of the bacteria to be used as starter cultures [10]). These criteria include tolerance to low pH and high ethanol and SO_2 concentrations, good growth characteristics under the winemaking conditions, compatibility with the *Saccharomyces cerevisiae* yeast used in AF, ability to survive the production process, inability to produce biogenic amines, lack of off-flavor or off-odor production, and the production of aroma compounds that may potentially contribute to a favorable wine aroma profile [11]. In the market, different companies commercialize various types of LAB starter cultures, which differ in their characteristics and how they are prepared before being added to the wines [11]. Between them, the freeze-dried direct inoculation cultures are the most useful, despite their expensiveness, because they do not need any special preparation, and they are directly added to the wine.

The timing of the inoculation of starter cultures is an essential factor influencing the success of induced fermentations. Several studies have been carried out to determine the effect of bacterial inoculation time on vinification kinetics, chemical composition, and sensory and sanitary attributes of wines [3, 12-15]. Results from some of these studies have shown that simultaneous yeast/bacteria inoculation poses important risks, such as the development of undesirable/antagonistic interactions between the two microorganisms, stuck AF, interruption of AF before sugar depletion, wines with increased concentrations of acetic acid that render them unacceptable for consumption, or the production of possible off-odors. Consequently, simultaneous inoculation has not been a prevalent practice, and the wineries have primarily adopted the addition of bacterial starter culture after the completion of AF. On the contrary, other authors [16] have recommended simultaneous addition of the yeast and the bacteria to the must, based on better performance of the bacteria, due to the low alcohol concentration and the higher nutrient availability present in musts, and to better sensory characteristics of the wines. Studies carried out with different grape varieties [17-20] have reported a reduction in total fermentation time and better control of the MLF, due to the early dominance of the inoculated bacterial strain.

In this chapter, the focus is on the effects of selected and commercial LAB on the chemical and sensory properties of Tempranillo wines. This grape variety is the most important red grape cultivar in Spain, predominantly in the Castilla-La Mancha winegrowing region. We are showing the results obtained about the behavior of an autochthonous *Oenococcus oeni* strain (named as C22L9), previously isolated and characterized at our laboratory [21], in comparison with two *O. oeni* commercial strains (PN4™ and Alpha™). Also, we will discuss how different LAB inoculation strategies can modulate the yeast and bacteria interactions and kinetics, the MLF time, and the chemical composition and sensory characteristics of wines.

EXPERIMENTAL DESIGN

In order to know the behavior of different *O. oeni* strains and the influence of the inoculation time of LAB (SEQ inoculation *versus* co-inoculation) on malolactic fermentation and on chemical composition and sensory quality of wines, two assays were carried out. Two commercial *Saccharomyces cerevisiae* strains, VRB™ and UvafermVN™, and three *O. oeni*, an autochthonous strain (C22L9) selected at our laboratory [21], and two commercial direct inoculation strains, PN4™ and Alpha™ were used. All of them supplied by Lallemand Inc. (Montreal, Canada).

Microvinification Assays

Must of Tempranillo grape variety from la Mancha wine region (Spain) was used. The chemical composition of the must was as follows: °Brix 23.60; total acidity 5.35 g/L tartaric acid; pH 3.42; L-malic acid 2.91 g/L; citric acid 0.33 g/L. Tempranillo must was added with SO_2 at an adequate concentration (50 mg/L when sequential inoculation was used and 40 mg/L for co-inoculation assays).

Microvinifications were carried out in an experimental cellar at the Vine and Wine Institute of Castilla-La Mancha (IVICAM). All fermentations were performed in triplicate. The progress of AF was monitored through glucose+fructose content (G+F), and for MLF, the L-malic acid and L-lactic acid content of the wines were determined.

In the first assay, traditional vinifications were carried out using the commercial yeast UvafermVN® for the alcoholic fermentation. After this process, the wine was racked and inoculated with one of the *O. oeni* assayed, allowing the MLF: strain C22L9 or the commercial strains,

PN4™ and Alpha™. Both fermentations occurred at 22 ± 2°C. In the second assay, the yeasts, *S. cerevisiae* VRB or *S. cerevisiae* VN, and the malolactic bacteria *O. oeni* C22L9 were used. *O. oeni* C22L9 was inoculated either after the completion of AF when glucose+fructose (G+F) content was below 1 g/L (SEQ), or just 24 h after yeast inoculation (COI), when free SO_2 concentration was less than 10 mg/L. When the L-malic acid content reached values ≤ 0.2 g/L, the wines were decanted and sulfated to reach a free SO_2 concentration of 25.0 mg/L and, subsequently clarified, stabilized, and filtered through 0.2 μm filters, following standard procedures before bottling.

Exhaustive chemical analysis, including major and minor volatile compounds, and sensory analysis of the resulting wines, were carried out. For the chemical analysis, the official analytical methods established by the International Organisation of Vine and Wine [22] were used, and the procedure described by Izquierdo-Cañas et al. (2020) [15] for the analysis of the volatile compounds. Sensory analysis, both descriptive and triangular tests, were carried out according to ISO Standard, 11035 (1994), and ISO Standard 4120 (1983), respectively. The Student-Newman-Keuls test for multiple comparisons of the means and multivariate data analysis (PCA) was used for the statistical analysis of the results, using SPSS 12.0 software (IBM, USA).

BEHAVIOR DURING MALOLACTIC FERMENTATION OF DIFFERENT *O. OENI* STRAINS

Figure 1 shows the evolution of the L-malic acid content in Tempranillo wines following inoculation with each of the three strains of *O. oeni* assayed [23]. Between 12 and 16 days were necessary to reach a L-malic acid content lower than 0.2 g/L. In all cases, the degradation of L-malic acid was prolonged during the first days following the inoculation. This fact has already been reported by other authors [24] and has been attributed to the characteristic of wine, such as the pH and the alcohol and

SO_2 contents, which make wine a very harsh environment for bacterial growth [7]. Commercial strains, PN4™ and Alpha™, required a somewhat more extended period (between one and three days) to consume the L-malic acid compared to the autochthonous C22L9 strain [23].

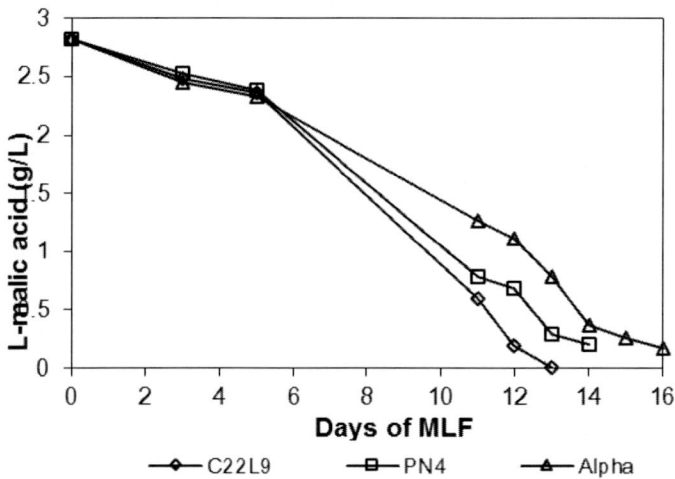

Figure 1. Evolution of L-malic acid content in wines inoculated with the *O. oeni* strains C22L9, PN4™ and Alpha™.

The malolactic activity of bacteria imparts recognizable changes to the flavor characteristics of the wine. These modifications come from the biotransformation and enzymatic processing of grape nutrients into flavor-active compounds [25] Several studies have shown that the combination of selected wine bacteria strains can modulate the amounts of specific metabolites, changing the aroma and flavor profile of Tempranillo wines [3, 15, 23]. Results for chemical parameters and volatile compounds most closely related to MLF are shown in Table 1 [23]. A decrease in the total acidity of between 0.79 to 1.08 g/L was observed in all wines following MLF, and no significant differences were obtained for the different strains of *O. oeni* used.

Consequently, an increase in the pH was obtained, ranging between 0.12 and 0.16 units for the two commercial strains, and 0.48 units for the C22L9 strain. The considerable increase in the pH of wines with the

C22L9 strain could be partially attributable to a slightly higher production of L-lactic acid and lower production of volatile acidity. However, other factors, such as the formation of organic acid salts [23], could also affect. The increase in the volatile acidity of the wines (between 0.01 and 0.12 g/L) was similar to that reported by Izquierdo-Cañas et al. (2015) [14].

Table 1. Chemical and volatile composition of wines inoculated with the strains of *O. oeni*, C22L9, PN4™ and Alpha™

	Before MLF	C22L9	PN4™	Alpha ™
Alcohol (% vol/vol)	13.84	13.73 ± 0.06	13.70 ± 0.16	13.66 ± 0.02
Total acidity (g/L)	5.22	4.28 ± 0.00	4.14 ± 0.24	4.43 ± 0.28
pH	3.57	4.05 ± 0.03[c]	3.73 ± 0.03[b]	3.69 ± 0.00[a]
Volatile acidity (g/L)	0.23	0.28 ± 0.02	0.30 ± 0.01	0.29 ± 0.01
L-malic acid (g/L)	2.82	0.01 ± 0.00[a]	0.20 ± 0.06[b]	0.17 ± 0.08[b]
L-lactic acid (g/L)	0.08	1.89 ± 0.03[c]	1.71 ± 0.01[a]	1.78 ± 0.08[b]
Citric acid (g/L)	0.32	0.28 ± 0.01[b]	0.15 ± 0.01[a]	0.24 ± 0.04[b]
2,3-Butanedione (mg/L)	1.83	3.54 ± 0.74[a]	9.23 ± 0.48[c]	7.02 ± 0.46[b]
3-Hidroxy-2-butanone (mg/L)	1.01	1.27 ± 0.09[a]	2.73 ± 0.02[c]	1.91 ± 0.07[b]
2,3- Butanediol (mg/L)	12.73	17.70 ± 6.43[b]	14.81 ± 0.22[a]	26.00 ± 4.46[b]
Acetaldehyde (mg/L)	13.69	4.17 ± 0.08[a]	4.84 ± 0.29[b]	5.47 ± 0.82[c]
Ethyl lactate (mg/L)	4.65	22.68 ± 1.15	22.45 ± 0.61	23.02 ± 1.43
Diethyl succinate (mg/L)	1.88	1.91 ± 0.05	1.76 ± 0.08	2.04 ± 0.33
Lineals alcohols (mg/L)	124.75	162.69 ± 5.24[b]	141.79 ± 1.82[a]	157.68 ± 16.30[b]
C6 alcohols (mg/L)	2.65	3.08 ± 0.13	3.05 ± 0.06	2.97 ± 0.02
Bencenic alcohols (mg/L)	27.20	25.11 ± 1.17[a]	24.52 ± 1.41[a]	27.85 ± 4.07[b]
Acids (mg/L)	5.18	5.98 ± 0.68	5.20 ± 0.68	6.15 ± 0.64
Acetates (mg/L)	1.82	1.65 ± 0.54	1.72 ± 0.18	1.97 ± 0.03
Ethyl esters (mg/L)	1.41	1.34 ± 0.06	1.34 ± 0.05	1.36 ± 0.06
Ethyl phenols (µg/L)	1.25	1.27 ± 0.07	1.23 ± 0.06	1.29 ± 0.01
Methoxyphenols (µg/L)	355	428 ± 107[a]	370 ± 66[a]	560 ± 35[b]
Terpenes (µg/L)	18	20 ± 1	18 ± 2	21 ± 2
Norisoprenoids (µg/L)	10	10 ± 1	9 ± 0	11 ± 2
Lactones (mg/L)	2.92	3.41 ± 0.83	3.09 ± 0.06	3.28 ± 0.38

Different letters (a, b, c) indicate significant differences between the *O. oeni* tested for α= 0.05 according to the test de Student-Newman-Keuls. Values are the mean of triplicates. The initial wine data were not statistically compared.

One of the most critical aromatic compounds produced by LAB in MLF is 2,3-butanedione, which at low concentrations (around 1.4 mg/L) contributes positively to the wine aroma, supplying buttery notes and adding complexity to the wine [26, 27], while at higher concentrations depreciates the quality. This compound is formed as an intermediate product in the metabolism of citric acid [28], which began when most of the L-malic acid has been transformed into L-lactic acid and, for this reason, the maximum concentration of 2,3-butanedione is reached when the L-malic acid is exhausted [26]).

It was observed that the degradation of citric acid and, consequently, the production of 2,3-butanedione and 3-hydroxy-2-butanone, was dependent on the strain of *O. oeni* used. The autochthonous strain C22L9 produced lower degradation of the citric acid than the two other strains, while the commercial strain PN4™ exhibited the highest degradation. However, the degradation of L-malic acid with the C22L9 strain was greater than that of the other two commercial strains.

Acetaldehyde is another critical compound associated with herbaceous and oxidative notes in wines [29]. In all cases, a decrease in acetaldehyde content was observed at the end of the MLF, and significant differences in the concentration of this compound were obtained for wines depending on the strain of *O. oeni* used. These results were in agreement with those reported by [30], who also observed differences in the final concentration of acetaldehyde in wines which MLF had been carried out with different *O. oeni* strains.

The esters most closely related to MLF are ethyl lactate and diethyl succinate [24, 31]. Ethyl lactate is one of the most important by-products of the metabolism of lactic acid bacteria, and it is beneficial for the wine aroma, supplying fruity and dairy notes, and contributing to the sensations of roundness in the mouth [24]. The concentration of ethyl lactate undergoes a significant increase following MLF, and although some authors [30] have reported that the concentrations reached are also dependent on the strain of *O. oeni*, in our study no significant differences were obtained. Diethyl succinate also contributes to the wine aroma, supplying fruity and melon notes [32]. In our study, contrarily to the reported by Ugliano et al. (2005), differences between strains were not statistically significant, although a higher content of this compound was observed in wines from strains C22L9 and Alpha™.

Linear alcohols contribute to the aromatic complexity of the wine, supplying a fruity flavor when they are found at concentrations lower than 300 mg/L. However, at concentrations above 400 mg/L, they are detrimental to the aroma [27]. During MLF, the linear alcohol content increased, and significant differences were observed between the strains, greater for the C22L9 and Alpha™ strains. These results are consistent with those obtained by Maicas et al. (1999) [33], who noted that the production of alcohols is dependent on the strain used to carry out MLF. Pozo-Bayón et al. (2005) [30] also observed increases in the alcohols during MLF, but statistically significant differences were not reported.

For other compounds, such as the C6 alcohols or the acids, no significant differences were observed between the three *O. oeni* assayed. It is worthnoting that the total acid concentration was less than 20 mg/L in all the wines, which does not compromise the quality or the aroma [30]. Concerning the acetates and ethyl esters, a disagreement between authors exists regarding the influence of MLF on the final content of these compound in wines, although some of them affirm that the production or hydrolysis of esters during MLF depends primarily on the LAB strain participating [11, 31]. In our study, the acetate and ethyl ester contents varied slightly during MLF, and increases or decreases were observed depending of the strain used, although for any of these compounds differences between strains were significant.

Results for methoxyphenols varied and despite the absence of significant differences in the wines' ethylphenols content before and after MLF with any of the strains studied, the increase observed for methoxyphenols during MLF was statistically higher for the strain Alpha™. In the case of terpenes, norisoprenoids, and lactones, volatile compounds were closely related to wine aroma [31], small changes in the content of these families of compounds were observed between the strains, although for some of them (i.e., terpenes) these differences were not statistically significant.

Principal component analysis (PCA) was applied to the results obtained from the chemical and volatile compound of the wines. Table 2 shows the variables with the highest correlation with principal component

1 (PC1) and principal component 2 (PC2). A total of 45.30% of the variance was explained by the first two principal components. Figure 2 shows the distribution of the wines on the plane formed by the two principal components PC1 and PC2. For PC1, two different groups were evident: the wines from strain PN4™, on the negative part of PC1, and those from the autochthonous strain C22L9 and the Alpha™ strain, located on the positive side of this axis. The latter had a higher content of propiovanillone, methyl vanillate, and benzyl alcohol.

Table 2. Results of principal component analysis (PCA) applied to the data from the chemical and volatile composition analyses in wines inoculated with the strains of *O. oeni* C22L9, PN4™ and Alpha™

Principal component	Variance explained %	Total Variance (%)	Variables highly correlated with the axis and their loadings
1	22.70	22.70	Propiovanillone (0.903)
			Methylvanillate (0.810)
			Benzyl alcohol (0.781)
			Damascenone (0.698)
			Acetosyringone (0.694)
			Diethyl succinate (0.679)
			2-Phenylethanol (0.648)
			L-lactic acid (0.619)
2	22.59	45.30	2,3-butanodione (0.851)
			Zingerone (0.817)
			pH (-0.803)
			3-hidroxy-2-butanone (0.783)
			Citric acid (-0.701)
			Ethanal (0.692)
			Volatile acidity (0.687)

Principal component 2 separated wines from PN4™ and Alpha™ strains from that of *O. oeni* C22L9, which was located on the negative side of this axis. Wines from *O. oeni* C22L9 had a lower content of 2,3-butanedione, 3-hydroxy-2-butanone and acetaldehyde, and a higher pH and citric acid content. Results from the triangular test carried for the sensory analyses of the pairs PN4™-Alpha™, C22L9-Alpha™, and C22L9-PN4™ showed significant differences only for wines from the C22L9 and PN4™ strains, with a 95% confidence interval. The wines elaborated with the autochthonous strain of *O. oeni* were preferred by 62.5% of the tasters when compared to the PN4™ wines.

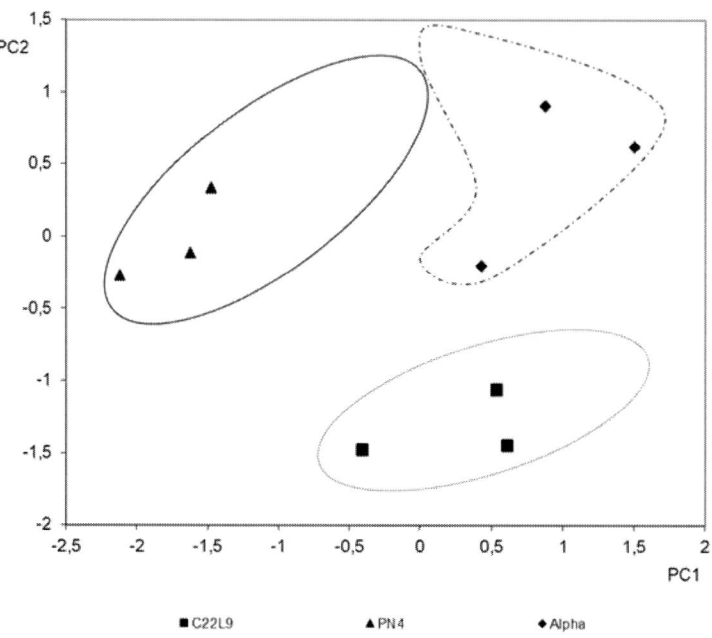

Figure 2. Distribution of the samples on the plane defined by the two principal component obtained by principal component analysis (PCA) of the data from chemical and volatile compound analyses of wines at the end of MLF with three strains of LAB (C22L9, PN4™ and Alpha™).

INFLUENCE OF THE *O. OENI* STRAIN INOCULATION TIME [SEQUENTIAL INOCULATION (SEQ) *VS* CO-INOCULATION (COI)] ON MALOLACTIC FERMENTATION

Synergies and antagonisms between microorganisms during winemaking are a common phenomenon in both the inoculated and non-inoculated LAB. These interactions are important when planning sequential inoculation or co-inoculation of LAB during winemaking [15]. Even though the biochemical bases for the synergies and antagonistic interactions between wine yeast and bacteria may be unclear, several factors could be involved [12].

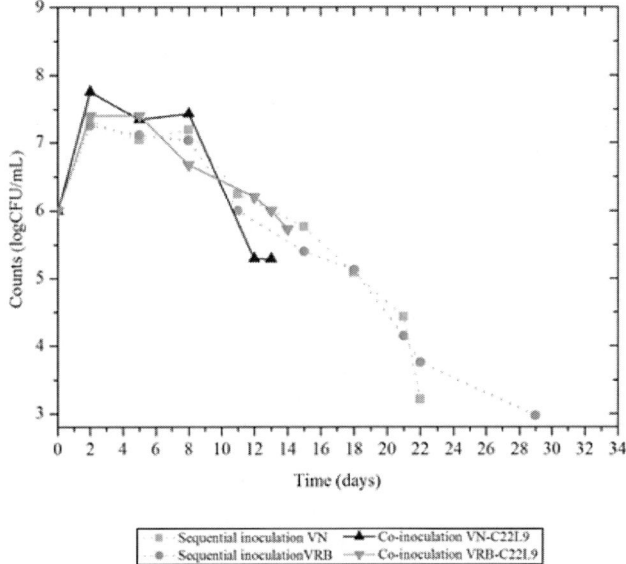

Figure 3. Yeasts counts from sequential inoculation and co-inoculation assays of VRB™ or VN™ yeasts and *O. oeni* C22L9. Values are mean of triplicates ± SE.

In our study, the behavior of the yeasts during the winemaking, *S. cerevisiae* VRB or VN, and the malolactic bacteria *O. oeni* C22L9 were analyzed. *O. oeni* C22L9 was inoculated either after the completion of AF when glucose+fructose (G+F) content was below 1 g/L (SEQ), or 24 h after yeast inoculation (COI), when free SO_2 concentration was less than 10 mg/L.

As reported by other authors [14, 18], the presence of *O. oeni* C22L9, during active AF in COI assays, did not affect the yeast population since the counts obtained from both inoculation time assays were similar (Figure 3). In contrast, evolution of LAB population (Figure 4) was different for SEQ and COI assays. In COI assays counts around 10^6 CFU/mL were obtained from the beginning of the fermentation and up to the end of AF, while in SEQ inoculation assays, counts of native LAB were lower until the inoculation of *O. oeni* C22L9 at day 8. It is important to highlight that in SEQ inoculation assays a slight decrease in LAB population was observed after inoculation, although at the end of MLF, LAB counts at both assays reached similar values.

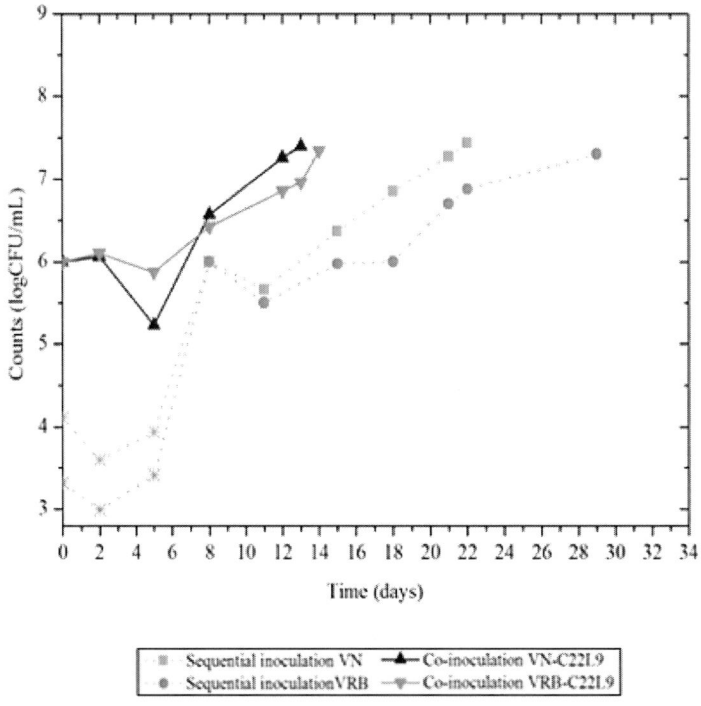

Figure 4. LAB counts from sequential inoculation and co-inoculation assays of VRB™ or VN™ yeasts and *O. oeni* C22L9. Values are mean of triplicates ± SE.

Results for G+F analysis in SEQ assays (Figure 5) showed almost complete depletion of these compounds, with final concentrations lower than 5 g/L. Values around 0.1 g/L of L-malic acid, the level generally recognized as the threshold for complete MLF, were reached some days earlier by the pair VN/C22L9. In contrast with the report by Massera et al. (2009) [18], results from L-malic acid analysis showed that native LAB presented in the SEQ assays before the inoculation of *O. oeni* C22L9 did not consume L-malic acid since its concentration remained constant until that moment. In COI assays (Figure 6) there was a slow decrease in L-malic acid content and an increase in L-lactic acid, almost immediately following the inoculation. From these results, it was concluded that the degradation of all these compounds (G+F and L-malic acid) was complete regardless of the time of the inoculation of LAB, in coincidence with results from Jussier et al. (2006) [17].

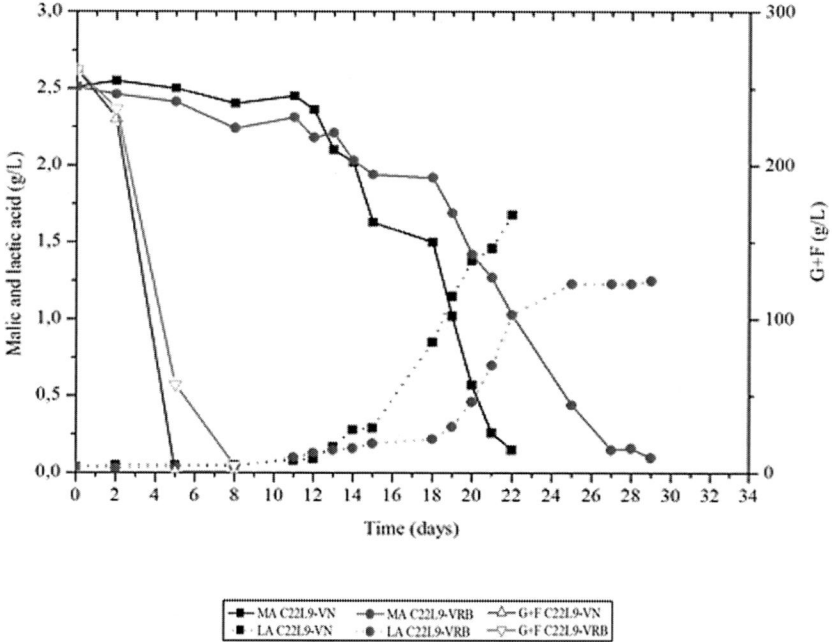

Figure 5. Time courses of glucose+fructose (G+F), L-malic (MA) and L-lactic acid (LA) concentrations during AF and MLF in wines obtained from sequential inoculation of VRB™ or VN™ yeasts and *O. oeni* C22L9. Values are mean of triplicates ± SE.

Major differences in overall fermentation (AF+MLF) duration were observed between SEQ and COI assays, with values ranging from 32 to 12 days, respectively. The time required for sugar concentration to fall below 1 g/L and L-malic acid concentration below 0.1 g/L was much shorter when COI was used. The length of MLF itself, measured as the time elapsing between LAB inoculation and depletion of L-malic acid, was longer in SEQ assays than the total fermentation time in COI treatments. These results are consistent with reports by some authors [17, 18, 20] and they are significant from a technological point of view because of the shorter time required to conclude the vinification process and the early microbiological stability conferred to wines. Moreover, these results were the same with both yeast strains assayed, VN, and VRB.

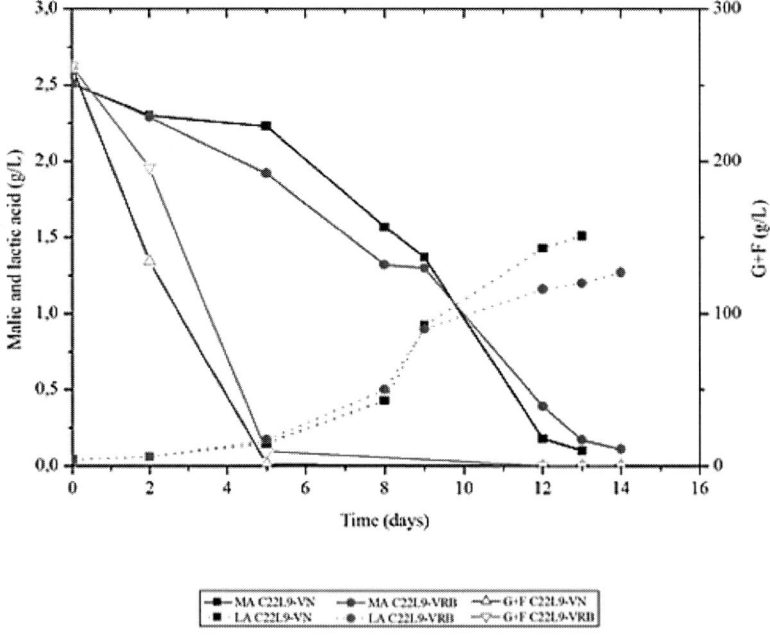

Figure 6. Time courses of glucose + fructose (G + F), L-malic (MA) and L-lactic acid (LA) concentrations during AF and MLF in wines obtained from co-inoculation of VRB™ or VN™ yeasts and *O. oeni* C22L9. Values are mean of triplicates ± SE.

Table 3 summarises the mean values±standard deviation of the chemical parameters and volatile compounds analyzed in the wines obtained from sequential and co-inoculation of VRB™ or VN™ yeasts and *O. oeni* C22L9 at the end of MLF. Values for volatile acidity, expressed as acetic acid, ranged from 0.30 to 0.51 g/L, which are in accordance to the standard quality parameter for volatile acidity in the red table wine [14]. Statistically higher values for volatile acidity were obtained in wines produced by co-inoculation. Total acidity and L-lactic acid contents were higher in wines from co-inoculation assays, with statistically significant differences for L-lactic acid depending on the yeast used. Values for citric acid concentrations in wines from co-inoculation assays were significantly higher than those from sequential inoculation assays.

Table 3. Chemical and volatile composition of wines obtained from sequential inoculation and co-inoculation of VRB™ or VN™ yeasts and *O. oeni* C22L9

Yeast	VRB ™		VN ™	
Inoculation method	SEQ	COI	SEQ	COI
Alcohol (% v/v)	13.62 ± 0.16	13.78 ± 0.23	13.93 ± 0.16	13.56 ± 0.33
Total acidity (g/L)	3.03 ± 0.17	3.12 ± 0.04	3.08 ± 0.04	3.53 ± 0.01*
pH	4.23 ± 0.01	4.22 ± 0.03	4.28 ± 0.00	4.23 ± 0.03*
Volatile acidity (g/L)	0.30 ± 0.03	0.36 ± 0.01*	0.46 ± 0.00	0.51 ± 0.01*
L-malic acid (g/L)	0.11 ± 0.02	0.11 ± 0.00	0.15 ± 0.00	0.11 ± 0.01*
L-lactic acid (g/L)	1.25 ± 0.00	1.27 ± 0.02	1.51 ± 0.01	1.68 ± 0.03*
Citric acid (g/L)	0.07 ± 0.01	0.31 ± 0.01*	0.04 ± 0.04	0.32 ± 0.02*
Linear alcohols (mg/L)	387.17 ± 4.79	413.70 ± 19.03*	301.55 ± 4.13	364.76 ± 11.96*
C6 alcohols (mg/L)	3.35 ± 0.19	4.33 ± 0.46*	4.09 ± 0.26	4.59 ± 0.08*
Bencenic alcohols (mg/L)	15.11 ± 0.15	17.79 ± 2.05*	16.36 ± 0.28	25.88 ± 2.36*
Thioalcohols (mg/L)	0.55 ± 0.04	0.28 ± 0.14*	0.51 ± 0.03	0.57 ± 0.02*
Furans (ug/L)	52.58 ± 4.03	32.53 ± 15.06	47.02 ± 5.16	29.97 ± 2.88*
Acids (mg/L)	9.58 ± 0.54	10.19 ± 1.19	9.52 ± 0.99	11.90 ± 2.06
Aldehydes and ketones (mg/L)	5.00 ± 0.24	3.67 ± 0.52*	4.36 ± 1.54	2.59 ± 0.15*
Esters (mg/L)	30.35 ± 1.89	29.48 ± 2.52*	28.72 ± 2.65	25.06 ± 0.61*
Terpens (ug/L)	22.12 ± 1.18	32.88 ± 1.96*	23.79 ± 1.22	30.14 ± 3.54*
Volatile phenols (ug/L)	9.94 ± 0.82	9.24 ± 0.79	9.39 ± 1.62	7.82 ± 0.91
Vanillate derivates (ug/L)	64.87 ± 9.41	75.35 ± 5.40	73.41 ± 3.83	66.45 ± 6.11
Norisoprenoids (ug/L)	0.95 ± 0.05	1.26 ± 0.03*	1.07 ± 0.09	1.06 ± 0.06

* Denotes statistically significant differences ($p \leq 0.05$) between the different inoculation methods. Values are the mean of triplicates.

Concerning linear alcohols, C6 alcohols and bencenic alcohols, wines obtained by COI contained more propanol, isobutanol and isoamilic alcohols and less 1-pentanol and syringol, with significant differences in some of these compounds. Likewise, there were higher concentrations of 1-hexanol, c-3-hexen-1-ol and benzyl alcohol, compounds that contribute significantly to wine aroma [34]. The remaining compounds of these groups behaved differently depending on the yeast strain inoculated. Thioalcohols concentrations were significantly lower in COI wines inoculated with the VRB™ yeast strain. This fact could be connected with the lower concentration of SO_2 added to the musts at the outset of AF and it is extremely important from a sensory standpoint since at high concentrations of these compounds impart notes of boiled vegetables, onion, etc. [3]. Furfuryl alcohol and furaneol were significantly lower in COI wines. On the contrary, furfural concentrations were higher in COI wines, with significant differences when VRB™ yeast was used.

Concentrations of the acids analyzed varied depending on the type of inoculation (SEQ or COI) and also the yeast used. Concentrations of hexanoic and octanoic acids were higher in COI wines. Acids impart notes herbaceous and fruity, fatty or rancid to wine [35] and even in low concentrations, their presence makes a significant contribution to wine aroma due to their low perception threshold [36]. Wines produced by co-inoculation also contained higher concentrations of 2,3-butanodione and lower concentrations of 3-hydroxy-2-butanone, 3-hydroxy-2-pentanone, and 1-hydroxy-2-propanone. Concentrations of the most of these compounds differed significantly from those found in SEQ wines. Low concentrations of these compounds give the wine aromatic complexity, with buttery notes, contributing positively to the aroma and sensory quality.

Esters are important for determining wine aroma, and the presence of some short-chain esters, such as ethyl acetate, isobutyl acetate, isoamyl acetate and hexyl acetate, contributes imparting fruity flavors. Thus, ethyl acetate at concentrations lower than 100 mg/L provides fruity notes, but it is responsible for an undesirable solvent/nail varnish-like aroma when it is present at high concentrations. Others such as diethyl succinate and ethyl lactate are beneficial in that they impart fruity, buttery, and creamy notes to wines and contribute to mouthfeel [11, 31, 32] COI wines contained significantly higher concentrations of ethyl acetate, ethyl butyrate, ethyl lactate, and diethyl succinate. Concentrations of most of these compounds differed significantly from the concentrations in SEQ wines. On the contrary, benzyl acetate concentrations were significantly lower in COI wines.

The ability of *O. oeni* strains to release terpenes from glycosidic precursors has been described [37] with the degree to which the enzymatic hydrolysis takes place being dependent on the bacterial strain, the chemical structure of the substrate, and the growth phase of the bacteria [11]. The results showed that wines produced by COI, with both the strains of yeast, presented statistically higher concentrations of α-terpineol, citronellol and geraniol and statistically lower concentrations of linalool. In contrast with these results, Knoll et al. (2012) [38] reported for Riesling wine that the SEQ wines had a higher content of α-terpineol while COI wines had a higher content of linalool.

Among the volatile phenols, ethyl-phenols are particularly important because they contribute negatively to the final quality of the wine, being responsible for the 'phenolic', 'animal' and 'stable' off-odors found in red wines [39]. Our study showed significant differences in some cases, in the concentrations of some of these compounds in wines produced by COI or SEQ. For example, concentrations of 4-vinylphenol and 4-vinyl-guaiacol were lower in wines produced by COI. Among the compounds belonging to the vanillate derivates group, which give wines their spicy and smoked characteristics [40], it was worth noticing the results from zingerone, whose concentrations were significantly higher in the wines produced by COI. The behavior of the remaining compounds in this group was dependant on the yeast used. Norisoprenoids also have a significant influence on the sensory quality of wines [31], contributing to the fruity, floral, or spicy notes. Concentrations of some of these, such as 3-hydroxy-β-damascone and 3-oxo-7,8-dihydro-α-ionol, differed significantly although any trend was observed.

Our results from this study confirmed the findings of other authors [3, 14, 20] for different grape varieties, demonstrating the possibility of simultaneous induction of AF and MLF without a pronounced degradation of L-malic acid during AF or an excessive increase in volatile acidity. The exhaustive chemical analysis carried out revealed numerous differences regarding the volatile compounds.

However, the impact of these differences on sensory profiles was limited according to the results of descriptive sensory analysis (Figure 7). Both yeast/bacteria pairs showed good compatibility and a similar behavior during the process, though the pair VN™/C22L9 was slightly better. Tasters detected slight sensory differences, such as a higher astringency of Tempranillo wine produced by SEQ, and less body and a more intense spicy, vegetable, and dairy aroma in COI wines, which could influence the typicity of this variety of wines.

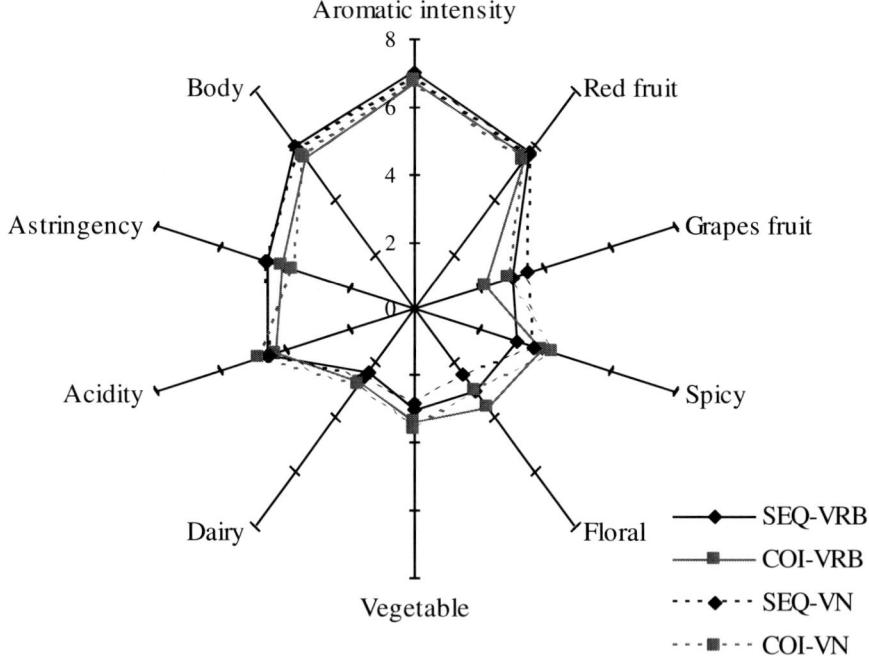

Figure 7. Descriptive sensory analysis of wines obtained from sequential inoculation and co-inoculation of VRB™ or VN™ yeasts and *O. oeni* C22L9.

Results reported above have displayed that both the LAB strain used for MLF, and the inoculation time influenced the chemical composition and sensory quality of wines and, therefore, both factors should be taken into account when a winemaking process is carried out.

REFERENCES

[1] Lonvaud-Funel, A. (1999). *Lactic acid bacteria in the quality improvement and depreciation of wine.* Antonie Van Leeuwenhoek, 76, 317-331.

[2] Ugliano, M., Genovese, A. & Moio, L. (2003). Hydrolysis of wine aroma precursors during malolactic fermentation with four commercial starter cultures of *Oenococcus oeni. Journal of Agricultural and Food Chemistry*, 51, 5073-5078.

[3] Izquierdo-Cañas, P. M., Pérez Martín, F., García Romero, E., Seseña Prieto, S. & Palop Herreros, M. L. (2012). Influence of inoculation time of an autochthonous selected malolactic bacterium on volatile and sensory profile of Tempranillo and Merlot wines. *International Journal of Food Microbiology*, 156, 245-254.

[4] Agouridis, N., Bekatorou, A., Nigam, P. & Kanellaki, M. (2005). Malolactic fermentation in wine with Lactobacillus casei cells immobilized on delignified cellulosic material. *Journal of Agriculture and Food Chemistry*, 53, 2546-2551.

[5] Izquierdo-Cañas, P. M., Ruíz Pérez, P., Seseña Prieto, S. & Palop Herreros, M. L. (2009). Ecological study of lactic acid microbiota isolated from Tempranillo wines of Castilla-La Mancha. *Journal of Bioscience and Bioengineering*, 108 (3), 220-224.

[6] Bauer, R. & Dicks, L. M. T. (2004). Control of malolactic fermentation in wine. A review. *South African Journal of Enology and Viticulture*, 25, 74-88.

[7] Coucheney, F., Desroche, N., Bou, M., Tourdot-Maréchal, R., Dulau, L. & Guzzo, J. (2005). A new approach for selection of *Oenococcus oeni* strains in order to produce malolactic starters. *International Journal of Food Microbiology*, 105, 463-470.

[8] Ribéreau-Gayon, P., Dubourdieu, D., Donèche, B. & Lonvaud-Funel, A. (2006). *Handbook of Enology: The Microbiology of Wine and Vinifications.* Chichester: Wiley and Sons.

[9] Izquierdo-Cañas, P. M., García E., Martínez, J. & Chacón, J. L. (2004). Selection of lactic bacteria to induce malolactic fermentation in red wine of cv. Cencibel. *Vitis*, 43, 149-153.

[10] Krieger-Weber, S. (2009). Application of yeast and bacteria as starter cultures. In H. Konig, G. Unden, & J. Frohlich (Eds.), *Biology of microorganisms on grapes, in must and in wine*, (pp. 498-511). Berlin: Springer.

[11] Lerm, E., Engelbrecht, L. & Du Toit, M. (2010). Malolactic fermentation: The ABC of MLF. *South African Journal of Enology and Viticulture*, 31 (2), 186-212.

[12] Alexandre, H., Costello, P. J., Remize, F., Guzzo, J. & Guilloux-Benatier, M. (2004). *Saccharomyces cerevisiae–Oenococcus oeni* interaction in wine: current knowledge and perspectives. *International Journal of Food Microbiology*, 93, 141-154.

[13] Abrahamse, C. E. & Bartowsky, E. J. (2012). Timing of MLF in Shiraz grape must and wine: influence on chemical composition. *World Journal of Microbiology and Biotechnology*, 28, 255-265.

[14] Izquierdo-Cañas, P. M., García Romero, E., Pérez Martín, F., Seseña Prieto, S. & Palop Herreros, M. L. (2015). Sequential inoculation versus coinoculation in Cabernet Franc wine fermentation. *Food Science and Technology International*, 21 (3), 203-212.

[15] Izquierdo-Cañas, P. M., Ríos-Carrasco, M., Garcia-Romero, E., Mena-Morales, A., Heras-Manso, J. M. & Cordero-Bueso, G. (2020). Co-existence of Inoculated Yeast and Lactic Acid Bacteria and Their Impact on the Aroma Profile and Sensory Traits of Tempranillo Red Wine. *Fermentation*, 6 (17), 1-13.

[16] Krieger, S., Zapparoli, G., Veneri, G., Tosi, E. & Vagnoli, P. (2007). Comparison between simultaneous and sequential alcoholic and malolactic fermentations for partially dried grapes in the production of Amarone style wine. *Australian & New Zealand Grapegrower & Winemaker*, 517, 71-77.

[17] Jussier, D., Morneau, A. D. & Mira de Orduña, R. (2006). Effect of simultaneous inoculation with yeast and bacteria on fermentation

kinetics and key wine parameters in cool climate Chardonnay. *Applied and Environmental Microbiology, 72*, 221-227.

[18] Massera, A., Soria, A., Catania, C., Krieger, S. & Combina, M. (2009). Simultaneous inoculation of Malbec (*Vitis vinifera*) musts with yeast and bacteria: effects on fermentation performance, sensory and sanitary attributes of wines. *Food Technology and Biotechnology, 47*, 192-201.

[19] Zapparoli, G., Tosi, E., Azzolini, M., Vagnoli, P. & Krieger, S. (2009). Bacterial inoculation strategies for the achievement of malolactic fermentation in high alcohol wines. *South African Journal of Enology and Viticulture, 30*, 49-55.

[20] Azzolini, M., Tosi, E., Vagnoli, P., Krieger, S. & Zapparoli, G. (2010). Evaluation of technological effects of yeast–bacterial co-inoculation in red table wine production. *Italian Journal of Food Science, 3* (22), 257-263.

[21] Ruiz, P., Izquierdo, P. M., Seseña, S. & Palop, M. L. (2010). Selection of autochthonous *Oenococcus oeni* strains according to their oenological properties and vinification results. *International Journal of Food Microbiology, 137*, 230-235.

[22] OIV. *International Organization of Vine and Wine.* (2019). OIV 2019 Reports.

[23] Izquierdo-Cañas, P. M., García Romero, E., Pérez Martín, F., Seseña Prieto, S., Heras Manso, J. M. & Palop Herreros, M. L. (2013). Behavior during malolactic fermentation of three strains of *Oenococcus oeni* used as direct inoculation and acclimatization cultures. *South African Journal of Enology and Viticulture, 34*, 1, 1-9.

[24] Ugliano, M. & Moio, L. (2005). Changes in the concentration of yeast-derived volatile compounds of red wine during malolactic fermentation with four commercial starter cultures of *Oenococcus oeni*. *Journal of Agricultural and Food Chemistry, 53*, 10134-10139.

[25] Styger, G., Prior, B. & Bauer, F. F. (2011). Wine flavor and aroma. *Journal of Industrial Microbiology & Biotechnology, 38*, 1145.

[26] Bartowsky, E. & Henschke, P. A. (2004). The "buttery" attribute of wine-diacetyl – desirability, spoilage and beyond. *International Journal of Food Microbiology*, 96, 325- 352.
[27] Swiegers, J. H., Bartowsky, E. J., Henschke, P. A. & Pretorius, I. S. (2005). Yeast and bacterial modulation of wine aroma and flavour. *Australian Journal of Grape and Wine Research*, 11, 139-173.
[28] Bartowsky, E., Costello, P. & Henschke, P. A. (2002). Management of malolactic fermentation-wine flavour manipulation. *Australian & New Zealand Grapegrower & Winemaker*, 461, 7-8 and 10-12.
[29] Osborne, J. P., Mira de Orduña, R., Pilone, J. G. & Liu, S. Q. (2000). Acetaldehyde metabolism by wine lactic acid bacteria. *FEMS Microbiology Letters*, 91, 51-55.
[30] Pozo-Bayon, M. A., G-Alegría, E., Polo, M. C., Tenorio, C., Martín-Álvarez, P. J., Calvo de la Banda, M. T., Ruiz-Larrea, F. & Moreno-Arribas, M. V. (2005). Wine volatile and amino acid composition after malolactic fermentation: effect of *Oenococcus oeni* and Lactobacillus plantarum starter cultures. *Journal of Agricultural and Food Chemistry*, 53, 8729-8735.
[31] Izquierdo-Cañas, P. M., García, E., Gómez, S. & Palop, M. L. (2008). Changes in the aromatic composition of Tempranillo wines during spontaneous malolactic fermentation. *Journal of Food Composition and Analysis*, 21, 724-730.
[32] Peinado, R. A., Moreno, J., Bueno, J. E., Moreno, J. A. & Mauricio, J. C. (2004). Comparative study of aromatic compounds in two young white wines subjected to pre-fermentative cryomaceration. *Food Chemistry*, 84, 589-590.
[33] Maicas, S., Gil, J. V., Pardo, I. & Ferrer, S. (1999). Improvement of volatile composition of wines by controlled addition of malolactic bacteria. *Food Research International*, 32, 491-496.
[34] Ugliano, M. & Henschke, P. A. (2008). Yeast and wine flavour. In: M. V. Moreno- Arribas, & C. Polo (Eds.), *Wine chemistry and biochemistry*, (pp. 328-348). New York: Springer.
[35] Mansfield, A. K., Schirle-Keller, J. P. & Reineccius, G. A. (2011). Identification of odor-impact compounds in red table wine produced

from Frontenac grapes. *American Journal of Enology and Viticulture*, 62 (2), 169-176.

[36] Rodríguez, S. B., Amberg, E. & Thornton, R. J. (1990). Malolactic fermentation in Chardonnay: growth and sensory effects of commercial strains of *Leuconostoc oenos*. *Journal of Applied Bacteriology*, 68, 139-144.

[37] Hernández-Orte, P., Cersosimo, M., Loscos, N., Cacho, J., García-Moruno, E. & Ferreira, V. (2009). Aroma development from non-floral grape precursors by wine lactic acid bacteria. *Food Research International*, 42, 773-781.

[38] Knoll, C., Fritsch, S., Schell, S., Grossmann, M., Krieger-Weber, S., du Toit, M. & Rauhut, D. (2012). Impact of different malolactic fermentation inoculation scenarios on Riesling wine aroma. *World Journal of Microbiology and Biotechnology*, 28, 1143-1153.

[39] Gerbaux, V., Briffox, C., Dumont, A. & Sibylle, K. (2009). Influence of inoculation with malolactic bacteria on volatile phenols in wines. *American Journal of Enology and Viticulture*, 60, 233-235.

[40] Ferreira, V., Fernández, P., Gracia, J. P. & Cacho, J. F. (1995). Identification of volatile constituents in wines from *Vitis vinifera* var. Viradillo and sensory contribution of the different wine flavour fraction. *Journal of the Science of Food and Agriculture*, 69, 229-310.

In: Fermented and Distilled ISBN: 978-1-53618-985-8
Editors: M. B. M. de Castilhos et al. © 2021 Nova Science Publishers, Inc.

Chapter 4

RED WINES: CARMENÈRE

Carolina Pavez[1,]*, Philippo Pszczólkowski[2],
Natalia Brossard[1] and Edmundo Bordeu[1]
[1]Department of Fruit Production and Enology,
Pontificia Universidad Católica de Chile, Santiago, Chile
[2]School of Agriculture,
Universidad Mayor, Santiago, Chile

ABSTRACT

Vitis vinifera cv. Carmenère is a red wine variety originating from the Bordeaux region of France. Carmenère was thought to be extinct after the phylloxera (*Daktulosphaira vitifoliae*) plague in Europe, but in 1994, it was rediscovered in Chile, where most of the worldwide vineyards are currently planted. From a viticultural point of view, Carmenère is a vigorous cultivar with a late bud break compared with other red wine varieties. Also, it is very sensitive to the condition of the soil where it is cultivated, and in general wheater conditions with high luminosity and daily thermal amplitude are better for its development. Carmenère is harvested quite late to avoid a high concentration of methoxypyrazines, which can produce undesirable vegetal aromas. However, canopy

---

* Corresponding Author's E-mail: capavezm@uc.cl.

management, in particular, leaf pulling is the best practice to handle this problem. Regarding winemaking practices, almost all are oriented to improve the final sensory quality (color and astringency) and diminish the methoxypyrazine content in wines. Carmenère red wine is deeply colored and exhibits well-structured tannins that produce a soft and round mouthfeel, which is expressed in a good level of astringency and a small degree of bitterness. On the other hand, Carmenère aroma profile is characterized by vegetal and herbaceous aromas primarily related to methoxypyrazines. Finally, Carmenère also has been characterized by fruity and spicy aromas that are related to C_{13}-norisoprenoids produced by hydrolysis of their glycosidic precursors, mostly during wine aging.

Keywords: Carmenère, red wine, late harvest, methoxypyrazines, color, astringency

ORIGIN AND PRINCIPAL CHARACTERISTICS OF CARMENÈRE RED WINE

Vitis vinifera cv. Carmenère is a red wine variety originating from the Bordeaux region, France. Carmenère was thought to be extinct after the phylloxera plague in Europe, but in 1994, it was rediscovered in Chile, where most of the worldwide vineyards of this grape variety are currently planted [1].

Nowadays, Carmenère is recognized as emblematic red wine variety in Chile due to its economic importance and international relevance [2]. Chile is by far the top producer of Carmenère grapes. According to the last viticultural report, performed by the Chilean Agricultural and Livestock Service (SAG) in 2018, Carmenère is the third most relevant red wine variety after Cabernet-Sauvignon and Merlot with an acreage of 10647 ha [3].

Regarding the viticultural behavior, Carmenère is a very vigorous cultivar that enters into production late, presenting low fertility in its basal buds. Another particularity is incomplete fertilization and reduced fruit set carrying bunches containing berries that considerably differ in size and maturity [2, 4].

According to the polyphenolic content, Carmenère is instensily colored and exhibits higher concentrations of glucoside and cumaroylglucoside anthocyanins throughout ripening, compared with other red wine varieties such as Cabernet Franc and Cabernet-Sauvignon [5]. Furthermore, Carmenère presents high concentrations and well-structured tannins generally with a soft and round mouthfeel, which confer to this red wine variety a great level of astringency and a small degree of bitterness [4, 6].

The aroma profiling of Carmenère is strongly dependent on the grape harvest. One of the primary features of Carmenère grape berries is their requirement for good sun exposure and a late harvest to enhance fruity aromas in detriment of vegetable nuances. However, this practice should be conducted carefully to avoid, for instance, the occurrence of fungal infection by *Botrytis spp* [1]. Another viticulture practice commonly used to improve fruity aromas is canopy management and leaf removing [7].

Vegetal and herbaceous aromas described as bell-pepper like are one of the most representatives of the Carmenère aroma profile. These aromas are primarily related to methoxypyrazine compounds (3-isobutyl-2-methoxypyrazine and 3-isopropyl-2-methoxypyrazine). Methoxypyrazines (MPs) are synthesized from the degradation of punctual amino acids by the grape berry metabolism. MP concentrations in Carmenère range between 2 and 45 ng/L. Additionally, it was also reported that MPs are strongly dependent on the ripening level of grapes, sun exposition, rain events, and terroir with lowest concentrations of these compounds described with higher sun exposition and over-ripening [8].

On the other hand, fruity and spicy aromas are mostly related to C_{13}-norisoprenoids produced by hydrolysis of their glycosidic precursors, mostly during wine aging. The most important C_{13}-norisoprenoids found in Carmenère are hydroxy- β-damascone, 3-oxo-α-ionol, and 3-hydroxy-β-ionone, among others. The concentration ranges of their glycosidic form are between 1500 to 7500 mg/L. The C_{13}-norisoprenoids compounds reach a maximum concentration close to grape ripening and they decrease with over-ripening [9].

Some varietal thiols, such as 2-furanylmethanethiol, 3-sulfanylhexyl acetate, 3-sulfanyl-1-hexanol, and 2-methyl-3-sulfanyl-1-butanol could influence the overall aroma in Carmenère [10]. This fact contradicts what happens with white wines since the role of varietal thiols has been widely studied. So far, the role of varietal thiols in red wines is not fully understood. However, they could be related to the enhancement of blackcurrant and fruity aromas, giving more complexity to the overall red wine aromas [11].

Viticulture and Enology: Influence on Carmenère Quality

Soil Conditions for Carmenère Culture

In Chile, Carmenère is still cultivated ungrafted due to the absence of *Phylloxera* and to the poor historical experiences with grafting with high vigor rootstocks in France during the time of the phylloxera crisis. Under this condition, Carmenère is very sensitive to limiting soil conditions. For example, soils high in clay with limited oxygen supply or the adverse conditions, sandy soils with low fertility, and limited water supply result in a rapid reduction of vigor and yields. Humid soils, in particular with a groundwater level close to the surface are not adequate because even if they induce higher vigor and yields, quality is strongly affected. Therefore, reliable drainage conditions are required. Additionally, good water supply is also relevant, soils with limitations in this sense because of textural problems or shallow, with impermeable strata are difficult to irrigate adequately resulting in low yields and limited growth [12].

Because an excess of vigor with consequences in a microclimate around bunches harms quality, fertilization and irrigation must be well implemented to obtain equilibrated plants. The use of some level of deficit irrigation has a positive impact on quality, promoting more sustainable viticulture, such as in other red varieties [13].

Climatic Conditions

Bud break of Carmenère is relatively late, between the dates of budbreak of Cabernet Franc and Cabernet-Sauvignon. Therefore, risks of spring frosts are relatively low. In general, this variety prefers climates with high luminosity and relatively high temperatures. In Chile, regions that present the preference for planting Carmenère correspond to those with high values of the MJT index (Mean January Temperature, the month with the highest temperature in the southern hemisphere) [14], such as Cachapoal and Colchagua valleys in O'Higgins region or areas like Pencahue in Maule region. Some of the best known Carmenère wines come from those areas [15]. Another relevant climatic factor for high-quality Carmenère wines is a high thermal amplitude between day and night, reflected in indexes such as the Fregoni index [16] and the cold night index proposed by Tonietto and Carbonneau (2004) [17].

High luminosity is a topic of particular interest in the case of Carmenère since reduces the intensity of pyrazine vegetal aromas, a punctual feature of Carmenère, which decreases the wine quality above some odorant threshold [18,19]. The degradation of these pyrazine aromas is one of the reasons that explain the postpone harvest dates performed by many of the Carmenère grape growers, so the absence of early fall rains is crucial to avoid risks of *Botrytis* rot. A high level of UV radiation, associated with this high luminosity stimulates higher levels of color and phenolic compounds such as anthocyanins, proanthocyanidins, and flavonols, which are relevant for quality and antioxidant properties.

For homogeneous bud break, Carmenère vineyards require a relatively high number of cold hours during winter [20]. The latter is something that limits its implantation in warm areas with a tropical or subtropical climates like Brazil, Uruguay, or Bolivia.

Production and Yields

In general, Carmenère produces relatively low yields and takes longer than other varieties to reach full production. This low production could be explained by the combination of two factors: low fertility of basal buds and fertility of flowers. Fertility, expressed as the number of bunches per bud left during pruning, is in general lower than Merlot and in particular very low in the two first buds of each cane (0,4 to 0,8 bunches per bud in southern Chile) [21].

Flowers suffer from abnormal stamens that have curved filaments and can be considered physiologically feminine [18]. This deformity of stamen filaments appears very early during flower development (Figure 1), and tese morphologic problems with flowers affect pollination, reducing fruit set. It gives a result in fewer berries and many smaller berries with fewer seeds and shot berries with no seeds. Typically, Carmenère berries are small and very diverse in diameter [21]. This condition is amplified with low temperature or rain during flowering and fruit set. Early bunch stem necrosis can be developed under these conditions affecting yields severely, an example of this is exhibited by some vineyards located in the south of Chile (Southern viticultural Region of Chile) [21].

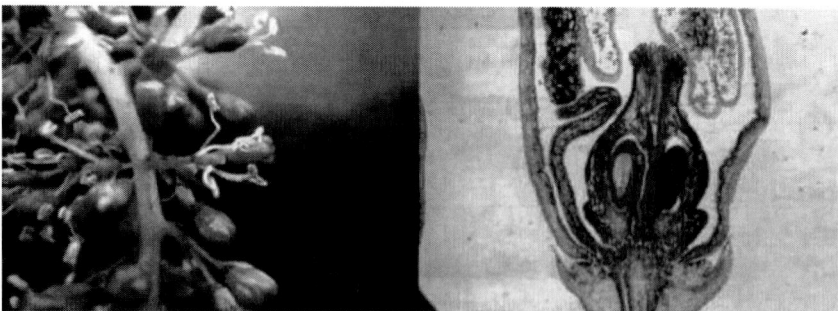

Figure 1. Abnormal Carmenère flowers characterized by stamen with curved filaments. This filaments are observed during early formation stage [22].

Maturation and Harvest

In the classical classification of varieties based on its relation with the harvest date of Chasselas, Carmenère is described with a ripening date only 3.5 weeks after Chasselas, similar to Cabernet-Sauvignon [18]. In Chile, however, the harvest date of Carmenère is much later, after Cabernet-Sauvignon, frequently during the second middle of April or even the first weeks of May, up to 100 days after *veraison*. Even more, some vineyards are harvested after almost complete leaf fall (Figure 2).

Acceptable sugar levels to obtain adequate levels of alcohol (from 13° to 14°) are acquired much earlier. Therefore, Carmenère harvested in its very late harvest tends to arrive to exaggerate levels of sugar that make fermentation very difficult and it tends to exaggerate the problem of low acidity and high pH, characteristic of Carmenère. The primary reasons to delay harvest are to improve phenolic maturity and reduce pyrazine aromas. In the case of Carmenère, only the second reason tend to apply.

Figure 2. Carmenère vineyard harvested after almost complete leaf fall. Maule Region, Chile.

Carmenère tannins produce wines with good mouthfeel and a velvety astringency, and, therefore, they do not require a harvest delay for this reason. Furthermore, the problem of pyrazine aromas is accentuated by cold weather conditions and shade microclimatic conditions determined by training systems such as the overhead horizontal trellis ("Parronal" in Chile) or vigorous vertical position trellising. Adequate trellising systems and canopy management are better viticultural solutions in most locations. These aromas are also variable season to season and it is still not clear that the delayed harvest date is always the best solution. Pszczólkowski and Henriquez (2002) [23] compared, with an expert panel of winemakers, wines from very different harvest dates (March, 23rd and April, 13rd). The results of this study showed that wines could not be significantly differed by the panel.

Carmenère Propagation

In Chile, the traditional way of propagation is mass selection simply collecting material from established vineyards during the pruning season. This simple and old-fashioned system of propagation has the well-known problems of high variability and risks of viral diseases and even mixed with other varieties in the case of material coming from old vineyards. The most typical mixture found in Chile is with Merlot, the variety that was wrong assigned to old Carmenère vineyards [24].

In Italy, there are a couple of clones of Carmenère ISV-FV5 and R9 that were confirmed as Carmenère by microsatellite analysis [18]. These two clones were used in the expansion of Carmenère to Canada and Brazil, even if they were originally classified as Cabernet Franc. In France, the clone 1059 is available. In Chile, there is a clonal selection program that started in 1999 at Pontificia Universidad Católica de Chile. The selection is based on aspects such as elimination of virus, grape and wine quality, and viticultural performance. A particular aspect considered for Carmenère has been looking for plants with normal flowers with straight stamens.

Carmenère is comparatively a variety that is relatively difficult to micro propagate.

In Chile, considering that Phylloxera is not present, almost all Carmenère vineyards are planted ungrafted. If somehow grafting is necessary, special attention must be paid to choose low vigor rootstocks to avoid some typical problems of this variety as the late start of production and in particular problems of fruit set, which have been considered to be one of the main reasons of Carmenère declination in Bordeaux after the phylloxera crises [25].

In Chile, Carmenère is generally planted in the classical vertical shoot positioning (VSP) trellis, inherited from the French tradition. Distances between rows are typically 2-2.5 m to allow relatively large equipment and 1 to 1.5 m on the row. Few vineyards are trained in the horizontal overhead system (parronal in Chile), frequently used for table grapes, 3.5-4 x 2-4 m, in general looking for higher yields. In the case of Carmenère, this trellising system is particularly inappropriate due to the difficulties in regulating vigor and shade. These conditions can promote a high concentration of pyrazines in grapes and, therefore, strong vegetal aromas. Additionally, with these conditions, it increases the risks of *Botrytis* with late harvest dates.

The shoots need two or more toppings to maintain the foliage with a height between 1.2 and 1.4 m, depending on the vigor of the plants. It is recommended to make these toppings early, cutting only the top of the shoots, where they are still growing. In this way, the amount of lateral shoots that could develop is strongly reduced [26]. This practice improves aeration and luminosity inside the canopy, both important for grape quality and sanity. Nowadays, most use mechanical topping machines. With a foliage height above 1.2 m, enough leaves are available for the ripening of bunches. An area of exposed leaves between 7 and 14 cm^2 per gram of grapes produced is adequate [15].

Another relevant practice that improves Carmenère quality is leaf pulling to enhance sun exposure of bunches. The sun exposition of bunches is the most useful practice to reduce the intensity of pyrazine aromas in wine. However, leaf pulling has to be conducted carefully to avoid grapes sunburnt. Especially in Chile, grape sunburnt is an increasing problem, apparently favored by the increase in UV radiation associated with the hole in the ozone layer. The best period to perform the leaf pulling to reduce the pyrazine content level in grapes is relatively early in the season. Scheiner et al. (2010) [27] found that the best dates to reduce pyrazine concentration were 10 and 40 days after flowering, while late treatments, after veraison, exhibited little effect. This leaf pulling of the bunched area is usually done by hand, but some specialized machines are being used in large vineyards.

Concerning the poor fruit set and yields of Carmenère, some studies of artificial pollination have been done with some improvement of fruit set bud. However, commercial applications have not developed yet. Aguilera and Pszczólkowski (2005) [28] compared pollination treatments with girdling the base of canes. The results showed better yields, without affecting wine quality in vineyards. Additionally, it was observed a good equilibrium between production and leaf area. The primary explanation of the significant yield increase is not due to more bunches production, but to the retention of more small berries without seeds. This effect is of particular interest in terms of wine quality due to the rapport skin to pulp increases (smaller average berry diameter). This higher proportion of anthocyanins and skin tannin is characteristic of Carmenère even if it is not girdled, and it is one of the explanations for the Carmenère wines intense color, good astringency quality, and mouthfeel.

Regarding pests, Carmenère is quite sensible to mites like *Brevipalpus chilensis* (false Chilean red mite) and in particular to *Colomerus Vitis* (grape blister mite). Considering Carmenère soil conditions sensibility, it is extremely sensitive to soil insects like phylloxera (*Daktulosphaira vitifoliae*), and *Margarodes vitis*, the same explains why Carmenère exhibits high sensitivity to nematodes. All this soil pests can reduce vigor and, in severe conditions, it can produce vineyard death [15].

On the other side, Carmenère is not particularly sensitive to the most relevant grapes diseases, mildew (*Erysiphe necator*), and *Botrytis cinerea*, despite being a late harvest variety. Late in autumn, the development of black fungus-like *Cladosporium herbarium* and *Cladosporium cladosporioides* on top of late and overripe berries is not very extended, which could be probably explained by the thick berry skin. Furthermore, late in the season, leaves become dark red, and the edge revolts towards the lower face, which could be attributed to a viral infection like leafroll or myoplasm [15]. Trunk disease problems are becoming increasingly important, like in other varieties, affecting seriously the vineyard longevity and uniformity.

Winemaking Techniques

Carmenère red wines are characterized by sensory attributes as high color content, soft tannins, and varietal methoxypyrazines (MP) aroma. Therefore, vinification practices are focused on achieving an optimal balance of these sensory characteristics.

Regarding the color content, Carmenère winemaking techniques are oriented in both the stabilization and proper polyphenol extraction. The color stability in red wines is given by several factors, such as the chemical structure of anthocyanins as monomeric pigments, pH of the must and wine, temperature, bisulfite concentration, oxygen concentration, and additionally due to the phenomenon known as copigmentation. The copigmentation in red wines consists of the formation of non-covalent complexes because of the structural affinity between pigments and other organic molecules, which in solution are generally not colored, such as anthocyanin glycosides, some phenolic acids, flavonoids, and in particular derivatives of flavonols and flavone subgroups. Copigmentation reactions in red wines can be seen as an intermediate step or a storage source for anthocyanins, stabilizing and protecting them from must/wine oxidation until polymeric pigments are formed [29-31].

Some enological practices can be proposed to achieve a high extractability and copigmentation: extended-maceration, *délestage*, co-winemaking, the addition of tannins, micro-oxygenation, wood aging, and aging on lees. Extended maceration and *délestage* (also called "rack and return"), are techniques that aim to obtain a good exchange between the must and pomace. *Délestage* is a maceration practice based on removing the liquid from the fermentation tank with consequent juice oxygenation, followed to a complete return of it into the pomace. The application of this practice promotes an increase of polymeric pigments, due to a high extraction of proanthocyanidins and oxygenation [32]. Extended-maceration can be used both before and after the period of fermentation, which has a differentiated effect on grape extraction. The pre-fermentative extension of the maceration allows a significant extraction of the monomeric anthocyanins that later polymerize and increase the color stability [33]. In contrast, extended maceration after fermentation involves the extraction of alcoholic soluble compounds as tannins that later can act as copigments [34]. However, due to the characteristic pyrazines profile and soft tannins exhibited by Carmenère, post-fermentative maceration is more commonly used than pre-fermentative.

Co-winemaking practice is based on the fact that certain grape varieties have more quantity of some volatile and non-volatile molecules. Therefore, maceration or fermentation using different grape varieties improves and balances the phenolic composition of the resulting wine [35,36]. With a similar objective, the external tannins addition technique is used, which can act as cofactors or source of copigments in conditions of copigment/pigment disbalance [37, 38].

Micro-oxygenation (MOX), wood aging, and aging on lees are conventional techniques with effects on red wine color stabilization. MOX involving the incorporation of small amounts of oxygen promotes reactions between anthocyanins and secondary metabolites [39-41]. MOX can be supplied during wine vinification or aging. However, this technique seems to have more significant influence before malolactic fermentation, when sulfur dioxide content is lower and monomeric forms of tannins and anthocyanins are present in higher concentrations [42]. On the other hand, the aging barrel allows controlling the oxygen incorporation as MOX. But at the same time, there is a transfer of hydrolyzable tannins due to the contact of the wine with oak barrels. This kind of tannins is important since they act as a cofactor in redox reactions that lead to color stabilization. Moreover, the use of alternative products to barrels (i.e., powder, chips, shavings, staves) in combination with MOX is also used to mimic the oak barrel stabilization process, but easier to implement and less expensive.

The lees consist of polysaccharides (also known as mannoproteins), nitrogen compounds, fatty acids, and nucleic acids released during yeast autolysis [32]. The benefits of aging on lees could be direct or indirect on wine. First, it has been demonstrated an influence on color stability by the interaction between anthocyanins and mannoproteins [43], and second, due to the capacity of lees to consume oxygen, they could protect the wine from both color and aroma oxidation [44].

For Carmenère aroma management, it is essential to highlight that the application of all practices mentioned above is depending primarily on the grape quality and their methoxy pyrazine (MP) content. First, in conditions of high levels of MPs, it is recommended to begin with a grape destemming [45] and to minimize pre-fermentative maceration. Second, related to fermentative processes, MPs are extracted from grapes, primarily in the first hours of skin-must contact. Therefore, maceration operations during fermentation certainly have a low effect on MPs extraction [46]. Third, aging technics as micro-oxygenation, oak barrels (or alternative products as chips, cubes, staves) and lees, it has been reported that their application reduces vegetative or green aromas. This is relevant to obtain Carmenère wines with a fruitier character or with a more complex profile [47]. Additionally, depending on the intensity of MPs on wines, blending with other grape varieties could be an alternative to mask the strong vegetative aromas of MPs. Finally, Carmenère grapes allow great versatility in winemaking techniques but require effective management of MP.

SENSORY CHARACTRISTICS: AROMA, TASTE AND COLOR

Principal Aromas and Sensory Descriptors

Every time that wine is consumed, it produces an interaction of taste, odor, and textural feeling, which provides an overall sensation defined as flavor. The flavor concept results from compounds responsible for taste and those responsible for odors. Compounds responsible for taste are generally nonvolatile at room temperature and they interact with the receptors located in the taste buds of the tongue. Meanwhile, aromas are volatile compounds that are perceived by odor receptor sites in the olfactory tissue of the nasal cavity [48].

Wine aroma is composed by around of 800 odorant compounds. Among them, one of the most interesting are those typical from the grape variety, such as monoterpenes and thiols, which have been found in Muscat and Sauvignon Blanc respectively [49,50]. Additionally, wine aromas can be distinguished according to their origin, i.e., those synthesized by the grape berry metabolism and those produced during winemaking [51].

Vitis vinifera cv. Carmenère has widely known for its vegetable, spicy and fruity odorant notes. The development of these aromas is due to the different odorant molecules produced among the whole winemaking process. The development of Carmenère aromas begins with the grape management in the vineyard until the aging process.

Varietal Aromas in Carmenère: Methoxypyrazines

Beginning with varietal aromas that characterize Carmenère red wines, one of the most important and widely studied are the family of methoxypyrazines (MP) [8]. MP are an important aroma impact compounds, and with a very low odor threshold (ng/L or 10^{-9} g/L) may influence the overall wine aroma [52]. MP are varietal compounds synthesized by the grape berry during early developmental stages. They reach a maximum concentration level before *veraison*. *Veraison* is the onset of the ripening and it is the moment when grape berries change their color. These compounds are responsible for the vegetal aromas, such as bell pepper and green beans [8].

In grapes, three types of MP have been identified: 3-isobutyl-2-methoxypyrazine (IBMP), 3-isopropyl-2-methoxypyrazine (IPMP), and 3-sec.butyl-2-methoxypyrazine (SBMP). Among them, IBMP has been reported as the most abundant with a low odor threshold in wine (15 ng/L). Concentration values of MP in wines are quite wide, ranging from 0.5 to 60 ng/L [53].

A study performed in 30 different Chilean Carmenère wines [8] found concentrations of IBMP ranging from 5.0 to 44.4 ng/L, which is far above the sensory detection threshold for this odorant compound (2 ng/L). The authors reported that MP in Carmenère red wines are strongly influenced by three factors as follows: vine genotype, climatic conditions, and viticultural practices.

For the vine genotype, it was reported wide differences among the Carmenère clones analyzed with a concentration of IBMP and IPMP ranging from 45 to 161 ng/L and 0 to 8.6 ng/L, respectively. These results suggest that the clonal selection of Carmenère vines is critical to managing the vegetal character of the resulting wine.

The climatic influence on Carmenère was studied for three years in three different viticultural Chilean valleys (Maipo, Cachapoal, and Colchagua). For this purpose, Carmenère grapes were monitored for four weeks before maturity and two weeks after maturity. Additionally, during the period of the experiment, it was addressed the thermal amplitude and rainfall. Results of this experiment verified that the major concentration of pyrazines was registered four weeks before maturity and then, MP concentrations were decreasing until reach a minimum at maturity. Interestingly, MP concentration for both IBMP and IPMP was similar in all valleys comparing the same harvest date. However, the parameter that strongly differentiates the MP concentration between different valleys was the sun exposition and rainfall, i.e., the valley with more sun exposition and less rainfall exhibited the lowest MP concentration. Therefore, as a conclusion, the climatic conditions, rather than terroir or maturity, present a critical effect upon MP in wines, especially on IBMP concentration.

Varietal Aromas in Carmenère: Thiols

Another relevant varietal aroma reported in Carmenère red wines is the family of varietal thiols [9, 10]. Thiol compounds are found as odorless precursors in grape berries linked to cysteine or glutathione. During alcoholic fermentation, the yeast releases them by the β-lyase enzymatic activity [51]. Thiol compounds have been widely studied on white wines and they are the primary compounds responsible for the exotic fruit odorant notes in Sauvignon Blanc wines. As the same as MP, thiols also exhibit an extremely low odor detection threshold, in the range of ng/L (10^{-9} g/L), in a synthetic wine model solution (10% v/v ethanol, pH 3.5). Therefore, due to this particularity, they are strong smelling compounds and, in quite small concentrations, they can influence the overall wine aroma.

The most widely volatile thiols found in wines are 3-sulfanyl-1-hexanol (3SH), 3-sulfanylhexyl acetate (3SHA), and 4-methyl-4-sulfanyl-2-pentanone (4MSP). These thiol compounds are associated with different sensory descriptors, such as grapefruit, passion fruit. Other relevant volatile thiols in wines are 2-furanylmethanethiol (FFT), and phenylmethanethiol (PhMT). FFT is responsible for roasted coffee or toasty aromas and it is primarily released by oaks barrels during aging [54]. PhMT is responsible for smoky aromas and its origin in wines is unclear [55]. Especially in Carmenère, it has been reported the presence of thiol compounds such as 3SH, 3SHA, FFT, and PhMT. So far, the role played by thiol compounds in the overall Carmenère aroma is still unclear. However, a study performed in other red wine varieties demonstrates a relationship between thiols concentrations in red wines and blackcurrant aroma intensity [56].

Additionally, the thiol compound 2-methyl-3-sulfanyl-1-butanol (2M3SB) has also been reported in Carmenère [10]. The compound 2M3SB has been identified in Sauternes wines, which are a type of wine produced in the Sauternes region of France from Sémillon, Sauvignon Blanc, and Muscat grapes colonized by noble rot (*Botrytis cinerea*) [57,58]. It is relevant considering that Carmenère grapes are harvested rather late to reduce the concentration of MP, which accounts for strong and undesirable vegetable-like aroma notes. The late harvest of Carmenère grapes matches the beginning of the rain season in Chile, which further promotes the occurrence of *Botrytis* as a result of high humidity [1].

Odor-Active Compounds in Carmenère

Among all volatile compounds present in Carmenère, only a small portion became odor-active compounds and, therefore, cause an impact in the overall perception of Carmenère. A useful tool to elucidate which volatile compound in a certain food matrix could become odor-active is the application of gas chromatography–olfactometry (GC–O). GC-O is a valuable technique to characterize odor-active, as well as character impact compounds, responsible for characterizing the odor of a food sample.

Furthermore, aroma extract dilution analysis (AEDA) is one of the most applied methods to determine the impact of the odor-active compounds in food samples. It consists of a GC-O analysis of the stepwise dilutions of the food extract until no odor perception at the sniffing port by the panelists. Odorants that reach the highest dilution factors (FD) are those with the major odor impact, and the main candidates to become key odor compounds. Results of AEDA are presented in an aromagram that correlates the resulting FD values and the retention indices of the odorants [59].

The solvent assisted flavor evaporation (SAFE), followed by the application of GC-O combined with AEDA analysis, identified twenty-one odor active zones with FD factors ≥ 16 (Table 1). The compounds with the highest FD factors were described as smelling fruity-sweet (1, 3, 8, 13, 14, 19), berry-like (12, 16), and spicy (7, 10, 11) [60]. The structural assignment of the odorants was carried out by comparing the retention indexes and odor qualities with a database [61]. Also, they were located two odorant zones, the first with a RI = 1426 described as green and earthy, and the other one with a RI = 1516 described as bell pepper. Comparing them with referenced compounds, they were identified as 3-isopropyl-2-methoxypyrazine and 3-isobutyl-2-methoxypyrazine, respectively. However, both odorant zones only reach a FD of 2.

From this study, the odor active compounds identified in Carmenère are primarily synthesized by the yeast metabolism during the alcoholic fermentation [62, 63]. However, there is an odorant zone with a retention index of 1007 and a dilution factor of 64 that is interesting to analyze. This odorant zone was identified as a mixture of methyl 2-methylbutanoate (M2MB) and methyl 3-methylbutanoate (M3MB). Both compounds are methyl ester and they could be formed in wine as a result of the reaction between carboxylic acids and methanol. The concentration of methanol in wines varies between 30 and 35 mg/L and resulted from the enzymatic hydrolysis of methoxy groups in the pectin of berry cell walls during fermentation [42]. So far, only a few methyl esters compounds have been reported in red wines. Compounds as methyl butanoate, methyl hexanoate, methyl octanoate, and methyl decanoate were found at concentration values close to 1 µg/L in Bordeaux red wines [64].

Both M2MB and M3MB were quantitated in 16 different red wines samples including another 8 red wines varieties as Cabernet-Sauvignon, Merlot, Malbec, Dornfelder, Shiraz, Nebbiolo, Tempranillo, and Pinot Noir [65]. The results showed that both methyl esters investigated were found in all samples analyzed in concentrations clearly below the sensory threshold value for M2MB (2.2 µg/L in model wine), and close or slightly above the threshold for M3MB (3.6 µg/L in model wine). Spiking experiments based on a synthetic red wine aroma model in combination with triangle tests did not show a significant aroma impact of M2MB and M3MB at the determined concentration levels.

Additionally, fruity and spicy aromas have been related to C_{13}-norisoprenoids released by hydrolysis from corresponding glycoside precursors. A study performed with Carmenère red wines from the Chilean viticultural valleys Maipo and Colchagua found the C_{13}-norisoprenoid compound β-damascenone with an odor activity value (OAV) close to 516 [9]. This OAV is an indicator that this compound is present well above its odor detection threshold and could be an important contributor to the Carmenère red wine aromas.

Table 1. Odor-active compounds (FD ≥ 16) in the SAFE distillate obtained from Carmenère red wine

n°	aroma compound [a]	odor quality [b]	RI [c] (FFAP)	FD [d]
1	2-/3-methylbutanol	fruity, sweet	1202	≥ 1024
2	2-phenylethanol	flowery, rose	1918	≥ 1024
3	ethyl butanoate	fruity	1029	256
4	acetic acid	vinegar-like	1435	256
5	2-/3-methylbutanoic acid	cheese	1667	256
6	5-butyl-4-methyldihidro-2(3H)-furanone	coconut	1959	256
7	3-hydroxy-4,5-dimethyl2(5H)-furanone	seasoning-like	2209	256
8	methyl 2-/3-methylbutanoate	fruity, sweet	1007	64
9	4-hydroxy-2,5-dimethyl-3(2H)-furanone	caramel, burnt sugar	2034	64
10	ethyl cinnamate	sweet, cinnamon	2132	64
11	4-allyl-2-methoxyphenol	clove-like	2159	64
12	ethyl 2-methylpropanoate	fruity, berries	947	32
13	ethyl 3-methylbutanoate	fruity, sweet	1066	32
14	ethyl hexanoate	fruity, sweet	1233	32
15	butanoic acid	cheese	1620	32
16	β-damascenone	plum, berries	1818	32
17	2-methoxyphenol	smoky	1863	32
18	benzyl alcohol	flowery	1877	32
19	ethyl 2-methylbutanoate	fruity	1045	16
20	3-(methylthio)propanal	cooked potato	1457	16
21	3-(methylthio)-1-propanol	cooked potato	1724	16

[a] The compounds were identified by comparing their retention indexes on capillary FFAP, mass spectra (MS-EI), and odor qualities as well as odor intensities as perceived during GC-O with the data obtained from the reference compounds under the same conditions.
[b] Odor quality perceived at the sniffing port.
[c] Retention index.
[d] Flavor dilution factor determined by AEDA on capillary FFAP.

Color and Astringency and Their Relationship with Wine Polyphenols

In food chemistry, both color and astringency are relevant sensory characteristics, which determine the final quality of a product and, as well as its consumer acceptance [66, 67]. In red wines and also in other plant-derived food products, the principal component responsible for color and astringency is the phenolic compounds [68].

Wine polyphenols have their origin in the grape berries, playing an important role in plant metabolism. Phenolic composition in grapes is highly affected by differences in grape varieties, environmental conditions, and viticultural practices [69]. Based on their chemical carbon skeleton, polyphenols are classified into two main families: flavonoids and non-flavonoids [68]. Grapes contain non-flavonoid compounds mainly in the pulp, while flavonoid compounds are located in the skins, seeds, and stem [70].

Non-flavonoids polyphenols consist of benzoic and cinnamic acids, while the group of flavonoids consists of flavanols (catechin monomers and oligomers), flavonols (quercetin), and anthocyanins [68, 71]. From the sensory point of view, flavanols can react to each other and polymerize originating the so-called tannins [71]. During red wine consumption, tannins interact with the salivary proteins producing a loss of lubricity inside the mouth, driving a textural sensation described as dry, rough, and puckering. This textural sensation is called astringency [72, 73].

On the other hand, anthocyanins are responsible for the red wine color. They are synthesized at the onset of *veraison* and they are accumulated in the vacuoles of the berry skin epidermal cells [71]. Structurally, they are glucosylated derivatives of cyanidin, peonidin, petunidin, delphinidin, and malvidin aglycones. Additionally, these aglycones could react with acetic and p-coumaric acid giving acetylated and coumaroylated derivatives. In grapes, the anthocyanin profile is characterized according to the variety and can be used as a distinguishing factor between different grapes varieties [68]. Further reactions of anthocyanins with proanthocyanidins contribute with color stabilization and they can influence and modify the astringency sensation [74, 75]. This sensory modification implies a reduction of the pucker sensation to becoming on a different astringent subquality, known as velvet astringency [75].

Characterization of Carmenère Grapes Polyphenols

Carmenère exhibits high concentrations of polyphenols due to the long maturation period exhibited by Carmenère grapes. This extended period of approximately 170 days after flowering promotes the polyphenol accumulation in the skin, and on the contrary, it reduces the polyphenols concentration in the seeds of grape berries [76].

The distribution of polyphenols in Carmenère grapes (Table 2) is primarily composed of anthocyanins in the skin (42%), while flavanols are concentrated in seeds (52%) [76-78]. Flavonols represent just a small proportion of Carmenère grape skin polyphenols (0.3%). However, it has been reported higher concentrations of flavonol compounds (6.5 and 2.4 mg/kg of quercetin and myricetin, respectively), which compared with other red wine varieties such as Merlot and Cabernet-Sauvignon is three times higher [76]. Furthermore, flavonol compounds have great importance to improve and stabilize the color in red wines; the latter due to flavonols that present the ability to interact with anthocyanins as a result of the co-pigmentation reactions [79].

Table 2. Polyphenolic distribution in Carmenère grapes

	Grape composition (%) *	
Compounds	Skin	Seed
Anthocyanins	42.3	nf
Flavonols	0.3	nf
Flavanols	0.3	52.0
Phenolic acids	0.1	5.0

* Percentage expressed from the total polyphenolic content reported in Carmenère grapes (adapted from Huaman-Castilla, et al. 2017 [76]).
nf: not found in seeds.

Carmenère is widely known to be a deeply colored variety, this distinctive characteristic is due to their high concentrations of anthocyanins [77]. So far, it has been identified 18 anthocyanins in Carmenère grape skin. Among them, it is possible to find monoglucoside anthocyanins and their acetyl, coumaroyl, caffeoyl, and feruloyl monoglucosides derivatives.

However, the major anthocyanin reported in Carmenère grapes is malvidin-3-*O*-glucoside, which exhibits high concentrations (862.2 mg/kg) in comparison with other red wine varieties such as Cabernet-Sauvignon and Merlot [76]. The importance of malvidin-3-*O*-glucoside is its role in the co-pigmentation that gives color stability to red wines [71]. Furthermore, by the interaction between malvidin-3-O-glucoside and other polyphenols, it has been observed an enhancement in the color intensity in aged red wines in about 30 and 50% [80].

Another important class of polyphenols present in Carmenère grapes is the family of condensed tannins or proanthocyanidins in monomeric and oligomeric form. The importance of these compounds is their ability to interact with salivary proteins and elicit the astringency sensation during wine consumption. In Carmenère grapes, the major proportion of condensed tannins is located in seeds (~90%) [81]. The most abundant monomeric forms are (+)-catechin and (-)-epicatechin. Additionally, it is possible to find dimers and trimers of proanthocyanidins such as epicatechin-(4β-8)-epicatechin, catechin-(4α-8)-catechin, catechin-(4α-8)-epicatechin-3-*O*-gallate, among others [77, 78, 81].

In comparison with Cabernet-Sauvignon grape seed tannin composition, Carmenère exhibits a lower content of monomeric flavan-3-ols, a higher mean degree of polymerization, a higher percentage of galloylation (~13.8%), a higher average of molecular weight in the flavonol fraction, a lower content of (+)-catechin, and a higher content of (-)-epicatechin, epicatechin-3-O-gallate, gallic acid, and galloylated procyanidin dimers [77,81].

On the other hand, the composition of tannins in Carmenère grape skin is considerably low comparing the tannin content in seeds. Among skin proanthocyanins, it has been reported (-)-epigallocatechin, (+)-gallocatechin, and (+)-catechin, as well as, proanthocyanidins with a high degree of polymerization. The latter are generally composed of sub-units of catechin and epigallocatechin monomers, with a low percentage of galloylation (~1.9%) [76].

Nevertheless, despite the high concentration of proanthocyanidins and even more, a high percentage of galloylation, the Carmenère variety has been described as soft and less astringent compared with Merlot and Cabernet-Sauvignon. An explanation of this fact could be the ratio between seed and skin proanthocyanidins. Normally, this ratio for Merlot and Cabernet-Sauvignon varies between 3 and 10. On the opposite, this ratio for Carmenère is around 2 [6]. Furthermore, Carmenère tannins are composed of more sub-units of epigallocatechin, which react in less extension with salivary proteins to form the aggregates that elicit the astringency sensation [82, 83].

Characterization of Carmenère Red Wine Polyphenols

The polyphenols found in Carmenère red wines are flavonols, anthocyanins, and flavan-3-ol derivatives molecules, such as quercetin and myricetin, malvidin, and epigallocatechin, respectively [76]. However, polyphenolic composition in red wines is strongly influenced by both viticultural and enological practices. For instance, on the viticultural side, the type of vineyard pruning, leaf removal, and the number of buds per plant are the most relevant practices that influence the final polyphenolic content in red wines [84, 85]. On the side of the enological practice, the most widespread techniques that have an important influence on the polyphenolic content are, for instance, fermentation temperature, thermovinification, must freezing, pectolytic enzyme treatment, extended maceration, and aging in oak barrels [29, 86].

So far, there is only a small quantity of research articles related to the polyphenolic composition in Carmenère wines [6, 79]. In one of these studies, 18 Carmenère wines from different enological Chilean valleys were analyzed according to their free flavonol composition after acidic hydrolysis. On average, Carmenère wines exhibit a high concentration of free flavonols as myricetin (12.3 mg/L) and quercetin (17.8 mg/L). In comparison with other red wine varieties (Cabernet-Sauvignon and

Merlot), Carmenère presented the higher mean concentration of the total of free flavonols [79].

The proanthocyanidin content in Carmenère wines presented catechin, epicatechin, epigallocatechin, and epicatechin gallate as extension units of the polymeric tannin composition, with the major concentration exhibited by epicatechin and epigallocatechin (56.9 and 35.5% in mol, respectively). About to another structural characteristic of proanthocyanidins, in average, Carmenère wines present a mean degree of polymerization value (mDP) of 7.4, a percentage of galloylation of 2.7, and the composition of the terminal polymeric units are primarily catechin and epicatechin (62.6 and 37.4% in mol, respectively) [6]. Overall, Carmenère wines exhibit more concentration of proanthocyanidins in comparison with Cabernet-Sauvignon, but despite this, Carmenère is less astringent. Nevertheless, Carmenère wine tannins are composed by epigallocatechin subunits, which has a poor interaction with salivary proteins [68] and, therefore, may modulate and diminish the Carmenère red wine astringency.

PERSPECTIVES ON THE DEVELOPMENT OF CARMENÈRE

The Carmenère variety was found in the Chilean vineyard in 1994, even though its minority existence was known in France and particularly in Italy, where it was confused with Cabernet Franc. Unlike these last two countries, Chile had the vision to invent the Carmenère concept, transforming the variety into an emblematic one of the country, and offering its wines to the rest of the world. Subsequently, the variety was also identified in vineyards in the Chinese province of Ningxia, which facilitated the penetration of Chilean wine in that market, where it currently occupies a second place, preceding traditional wine countries such as France, Italy, and Spain.

At the end of the 90's of the last century, an increasing number of investigations related to the variety was developed, particularly in Chile, both in the field of its agronomic management, as well as in its vinification and storage. Carmenère presents anatomical peculiarities in its flowers that affect its production levels, which ultimately affects its vegetative development. The latter must be carefully managed to avoid grapes from ripening in shady microclimates, a condition that leads to high levels of pyrazines.

Among the various aromatic compounds, it is precisely the high natural content of pyrazine compounds that significantly affect the sensory perception of Carmenère wines, which drove several investigations of how to make their winemaking, conservation, and storage, aiming at managing their final levels in this wine.

Currently, it is unthinkable to think of Chilean viticulture without the Carmenère variety and its wines drunk in multiple international markets.

REFERENCES

[1] Pszczólkowski, Ph. (2004). La invención del cv. Carménère (*Vitis vinifera L*) en Chile, desde la mirada de uno de sus actores. *Universum (Talca), 19*(2), 150-165.

[2] Gutiérrez-Gamboa, G., Díaz-Gálvez, I., & Moreno-Simunovic, Y. (2018). Effects of bud nodal position along the cane on bud fertility, yield component and bunch structure in 'carménère' grapevines. *Chilean Journal of Agricultural Research, 78*(4), 580-586.

[3] Servicio Agricola y Ganadero. (2018). *Catastro Vitivinícola Nacional.* Santiago, Chile. [*National Wine Registry.*]

[4] Gutiérrez-Gamboa, G., Liu, S. Y., & Pszczólkowski, Ph. (2020). Resurgence of minority and autochthonous grapevine varieties in South America: A review of their oenological potential. *Journal of the Science of Food and Agriculture, 100*, 465–482.

[5] Obreque-Slier, E., Peña-Neira, A., López-Solís, R., Cáceres-Mella, A., Toledo-Araya, H., & López-Rivera, A. (2013). Phenolic composition of skins from four Carmenère grape varieties (*Vitis*

vinifera L.) during ripening. *LWT - Food Science and Technology*, 54(2), 404–413.

[6] Fernández, K., Kennedy, J. A., & Agosin, E. (2007). Characterization of *Vitis vinifera* L. Cv. Carménère grape and wine proanthocyanidins. *Journal of Agricultural and Food Chemistry*, 55(9), 3675–3680.

[7] Fredes, C., Moreno, Y., Ortega, S., & Von Bennewitz, E. (2010). Vine balance: A study case in Carménère grapevines. *Ciencia e Investigación Agraria*, 37(1), 143–150. [*Agricultural Science and Research*]

[8] Belancic, A., & Agosin, E. (2007). Methoxypyrazines in grapes and wines of *Vitis vinifera* cv. Carmenère. *American Journal of Enology and Viticulture*, 4(58), 462–469.

[9] Domínguez, A. M., & Agosín, E. (2010). Gas Chromatography coupled with mass spectrometry detection for the volatile Profiling of *Vitis vinifera* Cv. Carménère wines. *Journal of the Chilean Chemical Society*, 3(55), 385–391.

[10] Pavez, C., Agosin, E., & Steinhaus, M. (2016). Odorant screening and quantitation of thiols in Carmenère red wine by gas chromatography-olfactometry and stable isotope dilution assays. *Journal of Agricultural and Food Chemistry*, 64(17), 3417–3421.

[11] Rigou, P., Triay, A., & Razungles, A. (2014). Influence of volatile thiols in the development of blackcurrant aroma in red wine. *Food Chemistry*, 142, 242–248.

[12] Pszczólkowski, Ph. (1997). El cv Carménère (*Vitis vinifera* L.), variedad peculiar del viñedo chileno. *Revista Frutícola*, 18(1), 27–30. [The Carménère cv (*Vitis vinifera* L.), a peculiar variety of the Chilean vineyard. *Fruit Magazine*]

[13] Gil, P., Knopp, D., Cea, D., Brossard, N., Zúñiga, A., & Bordeu, E. (2020). Innovación para el mejor uso del agua e insumos relacionados. In P. G. Montenegro & D. Knopp (Orgs.), *Acciones para una vitivinicultura sustentable e inocua*. Santiago: Facultad de Agronomía e Ingeniería Forestal, Pontificia Universidad Católica de Chile. [Innovation for the best use of water and related supplies. In P.

G. Montenegro & D. Knopp (Orgs.), *Actions for a sustainable and safe viticulture.*]

[14] Dry, P., & Coombe, B. (2005). *Viticulture. Volume 1: Resources* (2º ed, Vol. 1). Adelaide: Winetitles Media.

[15] Pszczolkowski, Ph. (2008). La culture du cépage Carmenère: L'optimum pour la qualité de son vin. *Le Progrès agricole et viticole, 125*(9), 163–172. [Cultivation of the Carmenère grape variety: The optimum for the quality of its wine. *Agricultural and viticultural progress*]

[16] Fregoni, M. (2014). *Viticoltura di qualità: Trattato dell'eccellenza da terroir.* Tecniche nuove. [*Quality viticulture: Treatise on excellence from terroir.*]

[17] Tonietto, J., & Carbonneau, A. (2004). A multicriteria climatic classification system for grape-growing regions worldwide. *Agricultural and Forest Meteorology, 124*(1–2), 81–97.

[18] Calo, A., Di Stefano, A., Costacurta, A., & Calo, G. (1991). Caratterizzazione de Cabernet et Carmenère (Vitis sp.) e chiarimenti sulla loro cultura in Italia. *Rivista di Viticoltura e di Enologia, 3*, 3–25. [Characterization of Cabernet et Carmenère (Vitis sp.) And clarifications on their culture in Italy. *Viticulture and Enology Magazine*]

[19] Belancic, A., Bordeu, E., & Agosín, E. (2005). Metoxipirazinas en Carmenère: Efecto del Terroir y fecha de cosecha. *Annals of the IX Congreso Latino Americano de Viticultura y Enología*, 63. [Methoxypyrazines in Carmenère: Terroir effect and harvest date. *Annals of the IX Latin American Congress of Viticulture and Oenology*]

[20] Dami, I., & Scurlock, D. (2005). Chilling requirements of Concord and Cabernet franc grapevine cuttings. *Annals of the 2005 ASEV Annual Meeting*, 54.

[21] Pszczólkowski, Ph. & González, A. (2005). *Diagnóstico de la problemática de baja producción en viñedos de uvas finas plantadas en la comuna de Coelemu, Trehuaco y Ranquil.* Facultad de Agronomía e Ingenieria Forestal, Pontificia Universidade Católica de

Chile, Santiago, Chile. [*Diagnosis of the problem of low production in vineyards of fine grapes planted in the commune of Coelemu, Trehuaco and Ranquil. Faculty of Agronomy and Forest Engineering*, Pontificia Universidade Católica de Chile, Santiago, Chile]

[22] Pszczólkowski, Ph., Pérez Harvey, J., Soto, A., Manssur, J., Montenegro, G., & Aguilera, I. (2007). Criterios de prospección implementados en Chile, tendientes a una selección clonal y sanitaria de vides Cv. Carmenère (*Vitis vinífera* L.). *Viticultura enología profesional*, *108*, 14–23. [Prospecting criteria implemented in Chile, aimed at a clonal and sanitary selection of Cv. Carmenère (*Vitis vinífera* L.). *Professional oenology viticulture*]

[23] Pszczólkowski, Ph., & Henríquez, I. (2002). Fecha óptima de cosecha del cv. Carménère. *Viticultura enología profesional*, *78*, 33–45. [Optimum harvest date of cv. Carménère. *Professional oenology viticulture*]

[24] Pszczólkowski, Ph. (2000). *Situación actual del material de propagación en Chile, ventajas y desventajas.* Facultad de Agronomía e Ingenieria Forestal, Pontificia Universidade Católica de Chile, Santiago, Chile. [*Current situation of propagation material in Chile, advantages and disadvantages.* Faculty of Agronomy and Forest Engineering, Pontificia Universidade Católica de Chile, Santiago, Chile.]

[25] Viala, P. & Vermorel, V. (1991). *Ampélographie*. Laffitte. [*Ampelography.*]

[26] Martínez de Toda, F. (1985). *Estudio de los efectos del despunte en la vid mediante utilización de radioisótopos* (PhD Thesis). Instituto de Estudios Riojanos, Madrid, Spain. [*Study of the effects of topping on the vine by using radioisotopes*]

[27] Scheiner, J., Sacks, G., Pan, B., Ennahli, S., Torlton, L., Wise, A., Lerch, S., & Heuvel, J. (2010). Impact of severity and timing of basal leaf removal on 3-isobutyl-2-methoxypyrazine concentrations in red winegrapes. American *Journal of Enology and Viticulture*, *61*(3), 358–364.

[28] Aguilera, I., & Pszczólkowski, Ph. (2005). Efecto del molibdato de sodio, polinización artificial y anillado de cargadores en la producción de cv. Carmenère (*Vitis vinífera* L). *Annals of the X Congresso Latino Americano de Viticultura e Enología*, 287. [Effect of sodium molybdate, artificial pollination and carrier ringing on cv production. Carmenère (*Vitis vinífera* L). *Annals of the X Latin American Congress of Viticulture and Oenology*]

[29] Sacchi, K. L., Bisson, L. F., & Adams, D. O. (2005). A Review of the Effect of Winemaking Techniques. *American Journal of Enology and Viticulture*, 48, 197–206.

[30] Boulton, R. (2001). The copigmentation of anthocyanins and its role in the color of red wine: A critical review. *American Journal of Enology and Viticulture*, 52(2), 67–87.

[31] Trouillas, P., Sancho-Garcia, J. C., De Freitas, V., Gierschner, J., Otyepka, M., & Dangles, O. (2016). Stabilizing and modulating color by copigmentation: Insights from theory and experiment. *Chemical Reviews*, 116(9), 4937–4982.

[32] Baiano, A., Scrocco, C., Sepielli, G., & Del Nobile, M. A. (2016). Wine processing: A critical review of physical, chemical, and sensory implications of innovative vinification procedures. *Critical Reviews in Food Science and Nutrition*, 56(14), 2391–2407.

[33] Auw, J. M., Blanco, V., O'keefe, S. F., & Sims, C. A. (1996). Effect of processing on the phenolics and color of Cabernet-Sauvignon, Chambourcin, and Noble wines and juices. *American Journal of Enology and Viticulture*, 47(3), 279–286.

[34] Cerpa-Calderón, F. K., & Kennedy, J. A. (2008). Berry integrity and extraction of skin and seed proanthocyanidins during red wine fermentation. *Journal of Agricultural and Food Chemistry*, 56(19), 9006–9014.

[35] Kovac, V., Alonso, E., & Revilla, E. (1995). The effect of adding supplementary quantities of seeds during fermentation on the phenolic composition of wines. *American Journal of Enology and Viticulture*, 46(3), 363–367.

[36] Gordillo, B., Cejudo-Bastante, M. J., Rodríguez-Pulido, F. J., Jara-Palacios, M. J., Ramírez-Pérez, P., González-Miret, M. L., & Heredia, F. J. (2014). Impact of adding white pomace to red grapes on the phenolic composition and color stability of Syrah wines from a warm climate. *Journal of Agricultural and Food Chemistry*, *62*(12), 2663–2671.

[37] Álvarez, I., Aleixandre, J. L., García, M. J., Lizama, V., & Aleixandre-Tudó, J. L. (2009). Effect of the prefermentative addition of copigments on the polyphenolic composition of Tempranillo wines after malolactic fermentation. *European Food Research and Technology*, *228*(4), 501–510.

[38] Baiano, A., Terracone, C., Gambacorta, G., & La Notte, E. (2009). Phenolic content and antioxidant activity of Primitivo wine: Comparison among winemaking technologies. *Journal of Food Science*, *74*(3), 258–267.

[39] Waterhouse, A. L., & Laurie, V. F. (2006). Oxidation of wine phenolics: A critical evaluation and hypotheses. *American Journal of Enology and Viticulture*, *57*(3), 306–313.

[40] Cejudo-Bastante, M. J., Pérez-Coello, M. S., & Hermosín-Gutiérrez, I. (2011). Effect of wine micro-oxygenation treatment and storage period on colour-related phenolics, volatile composition and sensory characteristics. *LWT - Food Science and Technology*, *44*(4), 866–847.

[41] Gómez-Plaza, E., & Cano-López, M. (2011). A review on micro-oxygenation of red wines: Claims, benefits and the underlying chemistry. *Food Chemistry*, *125*(4), 1131–1140.

[42] Ribéreau-Gayon, P., Glories, Y., Maujean, A., & Dubourdieu, D. (2000). *Handbook of enology. The chemistry of wine and stabilization and treatments* (2° ed, Vol. 2). Wiley.

[43] Gonçalves, F. J., Fernandes, P. A. R., Wessel, D. F., Cardoso, S. M., Rocha, S. M., & Coimbra, M. A. (2018). Interaction of wine mannoproteins and arabinogalactans with anthocyanins. *Food Chemistry*, *243*, 1–10.

[44] Rodríguez, M., Lezáun, J., Canals, R., Llaudy, M. C., Canals, J. M., & Zamora, F. (2005). Influence of the presence of the lees during oak ageing on colour and phenolic compounds composition of red wine. *Food Science and Technology International*, *11*(4), 289–295.

[45] Hashizume, K., & Samuta, T. (1997). Green odorants of grape cluster stem and their ability to cause a wine stemmy flavor. *Journal of Agricultural and Food Chemistry*, *45*(4), 1333–1337.

[46] Sala, C., Busto, O., Guasch, J., & Zamora, F. (2004). Influence of vine training and sunlight exposure on the 3-alkyl-2-methoxypyrazines content in musts and wines from the *Vitis vinifera* variety Cabernet-Sauvignon. *Journal of Agricultural and Food Chemistry*, *52*(11), 3492–3497.

[47] Sullivan, P. (2002). *The effects of microoxygenation on red wines* (PhD Thesis). California State University, Fresno, United States.

[48] Belitz, H. D., Grosch, W., & Schieberle, P. (2009). *Food chemistry* (4º ed). Springer.

[49] Schüttler, A., Friedel, M., Jung, R., Rauhut, D., & Darriet, P. (2015). Characterizing aromatic typicality of Riesling wines: Merging volatile compositional and sensory aspects. *Food Research International*, *69*, 26–37.

[50] Coetzee, C., Brand, J., Emerton, G., Jacobson, D., Silva Ferreira, A. C., & du Toit, W. J. (2015). Sensory interaction between 3-mercaptohexan-1-ol, 3-isobutyl-2-methoxypyrazine and oxidation-related compounds. *Australian Journal of Grape and Wine Research*, *21*(2), 179–188.

[51] Roland, A., Schneider, R., Razungles, A., & Cavelier, F. (2011). Varietal thiols in wine: Discovery, analysis and applications. *Chemical Reviews*, *111*(11), 7355–7376.

[52] Sidhu, D., Lund, J., Kotseridis, Y., & Saucier, C. (2015). Methoxypyrazine analysis and influence of viticultural and enological procedures on their levels in grapes, musts, and wines. *Critical Reviews in Food Science and Nutrition, 55*(4), 485–502.

[53] Kotseridis, Y., Baumes, R. L., Bertrand, A., & Skouroumounis, G. K. (1999). Quantitative determination of 2-methoxy-3-isobutylpyrazine

in red wines and grapes of Bordeaux using a stable isotope dilution assay. *Journal of Chromatography A*, *841*(2), 229–237.

[54] Blanchard, L., Tominaga, T., & Dubourdieu, D. (2001). Formation of furfurylthiol exhibiting a strong coffee aroma during oak barrel fermentation from furfural released by toasted staves. *Journal of Agricultural and Food Chemistry*, *49*(10), 4833–4835.

[55] Tominaga, T., Guimbertau, G., & Dubourdieu, D. (2003). Contribution of benzenemethanethiol to smoky aroma of certain *Vitis vinifera* L. wines. *Journal of Agricultural and Food Chemistry*, *51*, 1373–1376.

[56] Rigou, P., Triay, A., & Razungles, A. (2014). Influence of volatile thiols in the development of blackcurrant aroma in red wine. *Food Chemistry*, *142*, 242–248.

[57] Sarrazin, E., Shinkaruk, S., Tominaga, T., Bennetau, B., Frérot, E., & Dubourdieu, D. (2007). Odorous impact of volatile thiols on the aroma of young botrytized sweet wines: Identification and quantification of new sulfanyl alcohols. *Journal of Agricultural and Food Chemistry*, *55*(4), 1437–1444.

[58] Bailly, S., Jerkovic, V., Marchand-Brynaert, J., & Collin, S. (2006). Aroma extraction dilution analysis of Sauternes wines. Key role of polyfunctional thiols. *Journal of Agricultural and Food Chemistry*, *54*(19), 7227–7234.

[59] Guth, H., & Grosch, W. (1994). Identification of the character impact odorants of stewed beef juice by instrumental analyses and sensory studies. *Journal of Agricultural and Food Chemistry*, *42*(12), 2862–2866.

[60] Pavez Moreno, C. A. (2014). *Identification and characterization of odorant compounds in Carmenère red wine* (Ph.D. Thesis). Pontificia Universidad Católica de Chile, Santiago, Chile.

[61] Steinhaus, M., Sinuco, D., Polster, J., Osorio, C., & Schieberle, P. (2008). Characterization of the aroma-active compounds in pink guava (*Psidium guajava* L.) by application of the aroma extract dilution analysis. *Food Chemistry*, *56*(11), 4120–4127.

[62] Ferreira, V., Escudero, A., Campo, E., & Cacho, J. (2008). The chemical foundations of wine aroma–A role game aiming at wine quality, personality and varietal expression. *Proceedings of Thirteenth Australian Wine Industry Technical Conference*, 1–9.

[63] Frank, S., Wollmann, N., Schieberle, P., & Hofmann, T. (2011). Reconstitution of the flavor signature of Dornfelder red wine on the basis of the natural concentrations of its key aroma and taste compounds. *Journal of Agricultural and Food Chemistry*, *59*(16), 8866–8874.

[64] Antalick, G., Perello, M.-C., & de Revel, G. (2010). Development, validation and application of a specific method for the quantitative determination of wine esters by headspace-solid-phase microextraction-gas chromatography–mass spectrometry. *Food Chemistry*, *121*(4), 1236–1245.

[65] Pavez, C., Steinhaus, M., Casaubon, G., Schieberle, P., & Agosin, E. (2015). Identification, quantitation and sensory evaluation of methyl 2- and methyl 3-methylbutanoate in varietal red wines. *Australian Journal of Grape and Wine Research*, *21*(2), 189–193.

[66] Muñoz-González, C., Martín-Álvarez, P. J., Moreno-Arribas, M. V., & Pozo-Bayón, M. A. (2014). Impact of the nonvolatile wine matrix composition on the in vivo aroma release from wines. *Journal of Agricultural and Food Chemistry*, *62*(1), 66–73.

[67] Bucalossi, G., Fia, G., Dinnella, C., De Toffoli, A., Canuti, V., Zanoni, B., Servili, M., Pagliarini, E., Toschi, T. G., & Monteleone, E. (2020). Functional and sensory properties of phenolic compounds from unripe grapes in vegetable food prototypes. *Food Chemistry*, *315*, 126291.

[68] Cheynier, V., Dueñas-Paton, M., Salas, E., Maury, C., Souquet, J.-M., Sarni-Manchado, P., & Fulcrand, H. (2006). Structure and properties of wine pigments and tannins. *American Journal of Enology and Viticulture*, *57*(3), 298–305.

[69] Flamini, R., Mattivi, F., De Rosso, M., Arapitsas, P., & Bavaresco, L. (2013). Advanced knowledge of three important classes of grape

phenolics: Anthocyanins, stilbenes and flavonols. *International Journal of Molecular Sciences, 14*(10), 19651–19669.

[70] Monagas, M., Bartolomé, B., & Gómez-Cordovés, C. (2005). Updated knowledge about the presence of phenolic compounds in wine. *Critical Reviews in Food Science and Nutrition, 45*(2), 85–118.

[71] Nel, A. P. (2018). Tannins and anthocyanins: From their origin to wine analysis—A review. *South African Journal of Enology and Viticulture, 39*(1), 1–20.

[72] Soares, S., Brandão, E., Mateus, N., & de Freitas, V. (2017). Sensorial properties of red wine polyphenols: Astringency and bitterness. *Critical Reviews in Food Science and Nutrition, 57*(5), 937–948.

[73] Scollary, G. R., Pásti, G., Kállay, M., Blackman, J., & Clark, A. C. (2012). Astringency response of red wines: Potential role of molecular assembly. *Trends in Food Science &Technology, 27*(1), 25–36.

[74] Lorenzo, C., Pardo, F., Zalacain, A., Alonso, G. L., & Salinas, M. R. (2005). Effect of red grapes co-winemaking in polyphenols and color of wines. *Journal of Agricultural and Food Chemistry, 53*(19), 7609–7616.

[75] Gawel, R., Francis, L., & Waters, E. J. (2007). Statistical correlations between the in-mouth textural characteristics and the chemical composition of Shiraz wines. *Journal of Agricultural and Food Chemistry, 55*(7), 2683–2687.

[76] Huaman-Castilla, N. L., Mariotti-Celis, M. S., & Perez-Correa, J. R. (2017). Polyphenols of Carménère Grapes. *Mini-Reviews in Organic Chemistry, 14*(3), 176–186.

[77] Obreque-Slier, E., Peña-Neira, A., López-Solís, R., Zamora-Marín, F., Ricardo-Da Silva, J. M., & Laureano, O. (2010). Comparative study of the phenolic composition of seeds and skins from carménère and Cabernet-Sauvignon grape varieties (*Vitis vinifera* L.) during ripening. *Journal of Agricultural and Food Chemistry, 58*(6), 3591–3599.

[78] Obreque-Slier, E., López-Solís, R., Castro-Ulloa, L., Romero-Díaz, C., & Peña-Neira, A. (2012). Phenolic composition and physicochemical parameters of Carménère, Cabernet-Sauvignon, Merlot and Cabernet Franc grape seeds (*Vitis vinifera* L.) during ripening. *LWT - Food Science and Technology*, *48*(1), 134–141.

[79] Vergara, C., Von Baer, D., Mardones, C., Gutiérrez, L., Hermosín-Gutiérrez, I., & Castillo-Muñoz, N. (2011). Flavonol profiles for varietal differentiation between carmé nère and merlot wines produced in Chile: HPLC and chemometric analysis. *Journal of the Chilean Chemical Society*, *56*(4), 827–832.

[80] Lambert, S. G., Asenstorfer, R. E., Williamson, N. M., Iland, P. G., & Jones, G. P. (2011). Copigmentation between malvidin-3-glucoside and some wine constituents and its importance to colour expression in red wine. *Food Chemistry*, *125*(1), 106–115.

[81] Mattivi, F., Vrhovsek, U., Masuero, D., & Trainotti, D. (2009). Differences in the amount and structure of extractable skin and seed tannins amongst red grape varieties. *Australian Journal of Grape and Wine Research*, *15*(1), 27–35.

[82] Vidal, S., Francis, L., Guyot, S., Marnet, N., Kwiatkowski, M., Gawel, R., Cheynier, V., & Waters, E. J. (2003). The mouth-feel properties of grape and apple proanthocyanidins in a wine-like medium. *Journal of the Science of Food and Agriculture*, *83*(6), 564–573.

[83] Gibbins, H. L., & Carpenter, G. H. (2013). Alternative Mechanisms of Astringency—What is the Role of Saliva? *Journal of Texture Studies*, *44*(5), 364–375.

[84] Coletta, A., Trani, A., Faccia, M., Punzi, R., Dipalmo, T., Crupi, P., Antonacci, D., & Gambacorta, G. (2013). Influence of viticultural practices and winemaking technologies on phenolic composition and sensory characteristics of Negroamaro red wines. *International Journal of Food Science & Technology*, *48*(11), 2215–2227.

[85] Coletta, A., Berto, S., Crupi, P., Cravero, M. C., Tamborra, P., Antonacci, D., Daniele, P. G., & Prenesti, E. (2014). Effect of viticulture practices on concentration of polyphenolic compounds

and total antioxidant capacity of Southern Italy red wines. *Food Chemistry*, *152*, 467–474.

[86] Gambuti, A., Strollo, D., Erbaggio, A., Lecce, L., & Moio, L. (2007). Effect of winemaking practices on color indexes and selected bioactive phenolics of aglianico wine. *Journal of Food Science*, *72*(9), 623–628.

In: Fermented and Distilled
Editors: M. B. M. de Castilhos et al. © 2021 Nova Science Publishers, Inc.
ISBN: 978-1-53618-985-8

Chapter 5

TOURIGA NACIONAL RED GRAPE VARIETY: PHENOLIC AND AROMA COMPOSITION AND WINEMAKING TECHNOLOGY

Fernanda Cosme[*], *Luís Filipe-Ribeiro*
and Fernando Milheiro Nunes

CQ-VR, Chemistry Research Centre, Food and Wine Chemistry Lab.,
School of Life Sciences and Environment,
University of Trás-os-Montes and Alto Douro, Portugal

ABSTRACT

Touriga Nacional is one of the most important red wine grape varieties from Portugal, by its quality is being used at present in other countries, yielding red wines with outstanding quality. Touriga Nacional presents a complex yet elegant aroma, with fruit and floral sensory notes, sometimes with rose and violet flower character, with good color and phenolic composition allowing Touriga Nacional red wines present good aging characteristics. In order to deepen the knowledge of researchers, winemakers and consumers concerning Touriga Nacional grapes and

[*] Corresponding Author's E-mail: fcosme@utad.pt.

wines, this chapter describes the available data on the agronomical, enological and chemical characteristics of Touriga Nacional grape variety as well as the chemical and sensory characteristics of the wines, along with the influence of the *"Terroir"* on Touriga Nacional wine characteristics.

Keywords: Touriga Nacional, phenolic composition, aroma compounds, winemaking technology, *terroir*

INTRODUCTION

Touriga Nacional *Vitis vinifera* L. is a native and appreciated Portuguese grape variety, it is one of the main grape varieties cultivated in Portugal with a vine area of 13.032 ha spread throughout the Portuguese territory (Figure 1), representing 7% of the annual grape production [1].

Figure 1. Touriga Nacional red grape variety distribution in Portugal [2]. https://www.vinetowinecircle.com/en/castas_post/nacional-touriga/.

Figure 2. Bunch and leaf of Touriga Nacional [3].

Touriga Nacional is a vigorous grape variety with medium productivity and with a medium to early grape maturation period, considered as moderately sensitive to mildew, grey rot, powdery mildew, and susceptible to blight. This variety also presents good adaptability to different climatic conditions; nevertheless, it is very sensitive to excessive hot summers. Regarding the leaves, this variety contains adult leaves, quite heterogeneous, and with enormous polymorphism. The most characteristic leaves are medium to small, pentagonal in size, with five lobes with rectilinear short teeth, and a medium green color (Figure 2).

The bunch of this grape variety is small and moderately compact, containing berries with small to medium size, uniform with a rounded and short elliptical shape. Touriga Nacional grape berries have a black-blue color, thick film, non-colored, soft, and juicy pulp with an indefinite flavor. It requires good sunlight exposure for long periods, producing musts with good alcoholic potential and medium/high acidity [3].

The first references regarding Touriga Nacional grape variety cultivation were in 1790 by Lacerda Lobo, and in 1791 by Rebelo da Fonseca, who characterized this grape variety as "a productive and early maturing grape variety". Later, Visconde de Villa Maior, in 1865, praises Touriga, saying "it is excellent and gives very covered wine, resist to mildew". In 1900, Cincinato da Costa, in his monumental Portugal Vinímore, stated that Touriga Nacional was "a worthy red grape variety, generally appreciated throughout the north of the country for its high yields and superior quality of the wines", although in the Beira region and especially between the Mondego and Dão rivers, "the vineyards have a very characteristic imprint and justly famous, Touriga is the predominant grape variety". Touriga Nacional, first planted in the Dão region and later in the Douro Valley, is a recommended grape variety for Port wine production. It is in the Dão region, the believed origin of this variety, that Touriga Nacional vines present a greater genetic variability [4].

Touriga Nacional is also properly adapted to the edaphoclimatic conditions of the North-Eastern Brazilian region [5]. This grape variety shows a good adaptation to environmental stresses, withstanding high light intensities that allow for better adjustment to warm conditions, provided an adequate water supply [6].

Touriga Nacional produces wines presenting a high color intensity (closed ruby and violet hue), with a good aging potential, complex and intense aroma of red fruits (raspberry and cherry), black fruits (plum), wild berries (blackberry), jam, raisins and sometimes a very floral aroma (notably rose and violet) [7, 8]. This floral aroma, typical of Touriga Nacional, is due to the high content of free terpenoids, namely linalool, α-terpineol, nerol, and geraniol, with the violet aroma attributed to the high content of β-ionone [9, 10].

TOURIGA NACIONAL – GRAPE AND MUST COMPOSITION: EFFECT OF *"TERROIR"*

The chemical composition of Touriga Nacional grape berries and musts is shown in Table 1.

As can be observed in Table 1, Touriga Nacional grape berries and must show a wide range of values, related either to the harvesting year (Table 2) but also due to the different vitivinicultural *"Terroirs"* from each Touriga Nacional grapes produced in Portugal (Table 3). According to the International Organization of Vine and Wine (OIV) definition, vitivinicultural *"Terroir"* is a concept referring to an area in which collective knowledge of the interactions between the identifiable physical and biological environment and applied vitivinicultural practices develops, providing distinctive characteristics for the products originating from this area. *"Terroir"* includes specific soil, topography, climate, landscape characteristics, and biodiversity features" [14].

Table 1. Physicochemical composition of grape berries and must from Touriga Nacional grown in Portugal and Brazil at the technological maturity

	Portugal	Brazil	References
	Range of values		
Berry weight (g)	1.24 - 1.89		Fernandes (2009) [8]
Must volume (mL)[1]	142 - 136		Costa et al. (2015) [11]
Sugar content (g/L)	184.7 - 254.2		Petronilho (2015) [12]
Probable alchool content (% v/v)	11.7 - 14.0		Fernandes (2009) [8]
Total acidity (g tartaric acid/L)	3.78 - 7.28	4.0 - 6.7	Oliveira et al. (2018) [13]
pH	2.81 - 3.30	3.75 - 4.04	
Total polyphenol index (a.u.)[2]	57.6 -314		
Total anthocyanins (mg/L)	726 - 2 489		
Color intensity (a.u.)[2]	13.7 - 30.3	6.4 - 16.2	
Hue	0.56 - 0.73	0.49 - 0.63	

[1] Must volume extracted from 200 berries; [2] a.u. - absorbance units

The composition of the grapes and musts obtained from Touriga Nacional varies widely in the different Portuguese Demarcated Regions ([8], Table 3). Touriga Nacional grapes from the Vinhos Verdes region (Lousada), presented the highest berry weight value, as well as the lowest pH, which also corresponded to higher acidity. Touriga Nacional grapes from Dão and Vinhos Verdes regions presented the highest values in phenolic compounds (total polyphenol index - TPI), and total anthocyanins, and consequently also higher color intensity.

Table 2. Physicochemical composition of Touriga Nacional grape berries from the Bairrada Demarcated Region at technological maturity from the years 2010, 2011 and 2012 at harvests (Adapted from Petronilho, 2015 [12])

Parameters	2010	2011	2012
Berry weight (g)	1.6 - 1.9	1.6 - 1.7	1.7 - 2.0
pH	3.2 - 3.3	3.2 - 3.3	3.2 - 3.3
Acidity (g tartaric acid/L)	4.2 - 4.4	5.5 - 5.7	5.3 - 6.1
Sugar content (g/L)	199.4 - 254.2	193.8 - 205.1	184.7 - 201.2
Phenolic content (mg GAE/L)[1]	747.3 - 1341.3	948.9 - 1339.8	501.6 - 1121.4

[1] Folin and Ciocalteu

On the other hand, Touriga Nacional grapes produced in the Alentejo region (Cabeção and Vidigueira in the south of Portugal) presented the lowest values of berry weight, total acidity, as well as lower levels of phenolic compounds and total anthocyanins, and consequently lower values of color intensity.

According to Köppen climate classification, which characterizes the climate based on the average values of air temperature, the amount of precipitation and its correlated distribution over the months of the year, the climatic classification of Portugal mainland is classified as a mesothermal climate (C) with dry summers (s), designated as a Mediterranean climate (Cs) [15]. The Vinhos Verdes (Lousada), Dão (Carregal do Sal), and Lisboa (Leira) regions are classified more specifically as Csb, in which the average temperature of the hottest month is below 22 °C, but with an average temperature above 10 °C in at least 4 months of the year and, the Lisboa (Lisboa), and Alentejo (Cabeção and Vidigueira) regions, are Csa, which the temperature average of the hottest month is over 22 °C (Figure 3).

Also, according to the FAO soil classification, in the region of Vinhos Verdes, Dão, and Lisboa (Leiria), the soils are of the cambisolo type, the first two of granite origin and the last of limestone origin. The soils from the Lisboa region are classified as vertisols. In Alentejo, Cabeção is a Podzois soil, and finally in Vidigueira also in Alentejo, the soil belongs to the class of litossolos [16]. Therefore, these differences in grape and must composition observed for the Touriga Nacional grown in different Portuguese regions can be justified taking into account the optimal conditions for the anthocyanins accumulation in the grape berries, which are between 15 and 25 °C during the day [17], and the great thermal amplitudes (as in Cabeção and Vidigueira in Alentejo region), [18-20] decrease significantly the anthocyanin content in berries.

Also, significant differences in Touriga Nacional grapes and must at technological maturity were described between Dão and Douro region by Costa et al. (2015) [11], (Table 4).

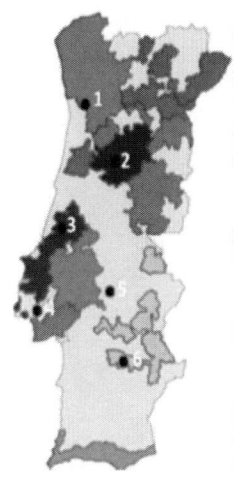

	Vinhos Verdes (Lousada)	Dão (Carregal do Sal)	Lisboa (Leiria)	Lisboa (Lisboa)	Alentejo (Cabeção)	Alentejo (Vidigueira)
Average annual rainfall (mm)	2000 - 2400	1000 - 1200	800 - 1000	500 - 600	500 - 600	500 - 600
Average annual rainfall in days	> 100	75 - 100	> 100	50 - 75	75 - 100	75 - 100
Annual average temperature (°C)	10 - 12.5	15 - 16	12.5 - 15	15 - 16	12.5 - 15	15 - 16
Relative humidity (%)	> 85	80 - 85	75 - 80	70 - 75	65 - 75	75 - 80
Frost in days	10 - 20	30 - 40	10 - 20	1 - 5	30 - 40	5 - 10
Real evapotranspiration (mm)	700 - 800	600 - 700	600 - 700	450 - 500	500 - 600	400 - 500
Real annual average insolation (h)	2300 - 2400	2600 - 2700	2300 - 2400	2800 - 2900	2700 - 2800	2900 - 3000
Solar radiation (kcal/cm^2)	140 - 145	140 - 145	140 - 145	150 - 155	145 - 150	150 - 155
Average annual surface runoff (mm)	1000 - 1400	600 - 800	200 - 300	50 - 100	25 - 50	150 - 200

Figure 3. Map of the different regions with Touriga Nacional and climatic data of each region. Adapted from Atlas do Ambiente Digital – Instituto do Ambiente (Adapted from Fernandes, 2009 [8]).

Table 3. Data from Touriga Nacional, weight per berry, probable alcohol content (PAC), pH, total acidity, total anthocyanins content, total polyphenol index (TPI) and color hue/intensity, at harvest date (Adapted from Fernandes, 2009 [8])

	Berry weight (g)	PAC (% v/v)	pH	Total acidity (g tartaric acid/L)	Total anthocyanins (mg/L)	TPI (a.u.)	Hue	Color intensity (a.u)
VinhosVerdes (Lousada)	1.89	11.7	3.31	7.3	1827.7	80.5	0.56	24.0
Dão (Carregal do Sal)	1.76	12.0	3.38	5.9	2489.2	104.4	0.58	30.3
Lisboa (Leiria)	1.68	13.1	3.45	5.7	1510.8	67.9	0.61	21.3
Lisboa (Lisboa)	1.30	13.1	3.40	4.2	1787.7	76.3	0.59	23.0
Alentejo (Cabeção)	1.36	12.2	3.85	3.9	990.8	60.3	0.73	13.7
Alentejo (Vidigueira)	1.24	13.5	3.57	4.1	1335.4	57.6	0.65	15.8

PAC – probable alcohol content (%), TPI – total polyphenol index.

Table 4. Physicochemical composition of grape must from Touriga Nacional studied at technological maturity (Adapted from Costa et al., 2015 [11])

	Total polyphenol index	Must volume (mL)	Estimated alcohol degree (% v/v)	pH	Titratable acidity (g/L tartaric acid)
Dão	254 ± 1	142 ± 4	12.91 ± 0.04	3.19 ± 0.01	3.78 ± 0.03
Douro	314 ± 2	136 ± 6	13.97 ± 0.01	2.81 ± 0.01	6.00 ± 0.01

The composition of Touriga Nacional grape berries and must grown in the tropical semi-arid northeast region of Brazil, it was observed differences between harvests seasons, shown the highest pH values (3.70 to 4.00) [13] when compared to the musts produced from Touriga Nacional grown in Portugal (Table 1). This fact can be related to the high potassium contents present in the soils of this region [21]. On the other hand, the total acidity values of the grapes grown in this Brazilian region (5.3 g/L in the harvest of 2016 and 6.7 g/L in the harvest of 2017 [13] are in the range of those described for grapes grown in Portugal (Table 1).

There is strong evidence that temperature affects grape berry composition, but only differences in 3°C or above are reported as causing changes [22, 23]. On the other hand, the biosynthesis of anthocyanins is sensitive to the environment light, and it decreases with lower light intensity [24, 25]. Shaded berries from Touriga Nacional had a lower concentration of total anthocyanins and extractable anthocyanins than non-shaded berries (Table 5).

The effect of surrounding vegetation height on Touriga Nacional grape variety, from the Douro Demarcated Region, in Portugal, was also evaluated. The results showed that grapes grown in vineyards with higher vegetation height (100 cm) had higher carotenoid contents, while grapes grown in vineyards with lower vegetation height (60 cm) with higher sunlight exposure, had higher grape berry weight and sugar content. Additionally, during the maturation period, a decrease in carotenoid degradation was observed in vineyards with higher vegetation height, explaining their higher carotenoid content [27].

Jordão and Correia (2012) [28] studied the evolution of anthocyanins in Touriga Nacional grape skins and procyanidins in Touriga Nacional grape skins and grape seeds during maturation. As expected, the content of anthocyanins in Touriga Nacional grape skins increased during the maturation, highlighting that, except malvidin-3-glucoside for the last two weeks, increased during maturation (Figure 4).

Table 5. Effect of shading on total anthocyanins and extractable anthocyanins for each shading treatment. *Vitis vinifera* cv Touriga Nacional grape musts, Alto Douro (Adapted from Oliveira et al., 2014 [26])

Shading	Total anthocyanins (mg/L)	Extractable anthocyanins (mg/L)
Non-shaded	1534^a	825^a
After fruit setting	990^b	679^b
After *véraison*	1240^b	726^b

Different superscript letters on same column indicate a significant difference at $\alpha \leq 0.05$ (Tukey's HSD test).

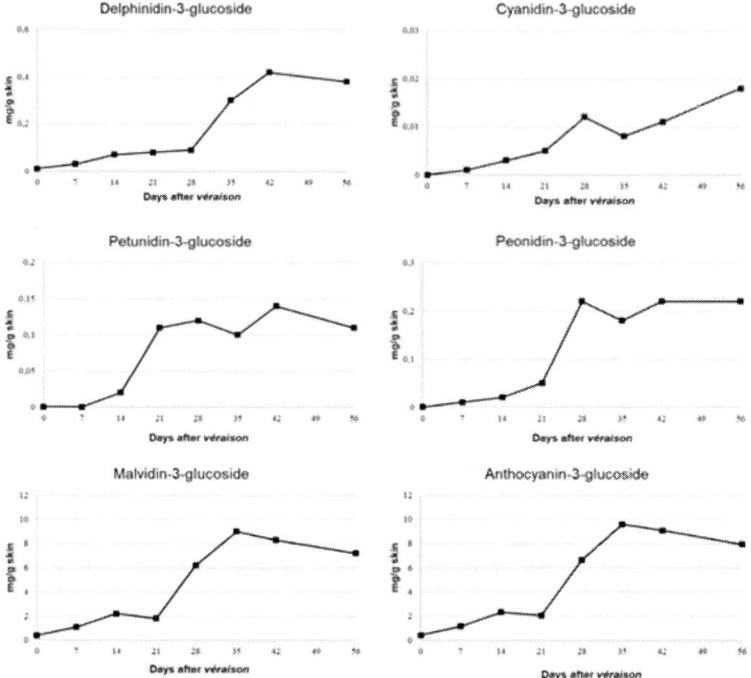

Figure 4. Evolution of skin anthocyanin glucoside derivatives during grape maturation of the Portuguese Touriga Nacional grape variety (Adapted from Jordão and Correia, 2012 [28]).

On the other hand, for procyanidins evolution during Touriga Nacional maturation, it was observed a decrease in the monomeric, oligomeric and polymeric procyanidins, except for the last two weeks when the concentration was maintained. During grape berry ripening, seeds presented the highest content in all proanthocyanidin fractions (56.0 mg/g of seeds at the beginning of *véraison* to 8.0 mg/g of seeds at the end of maturation for the oligomeric fraction) [28] (Figure 5). The decrease in proanthocyanidin concentration observed in the seeds after *véraison* can be attributed to the oxidation reactions [29]. On the other hand, Cheynier et al. (1997) [30] attributed this decrease to the reduced extractability resulting from the conjugation of proanthocyanidins with other cell components, and Valero et al. (1989) [31] considered this decrease in the concentration of procyanidins during grape maturation related to the increase in berry weight.

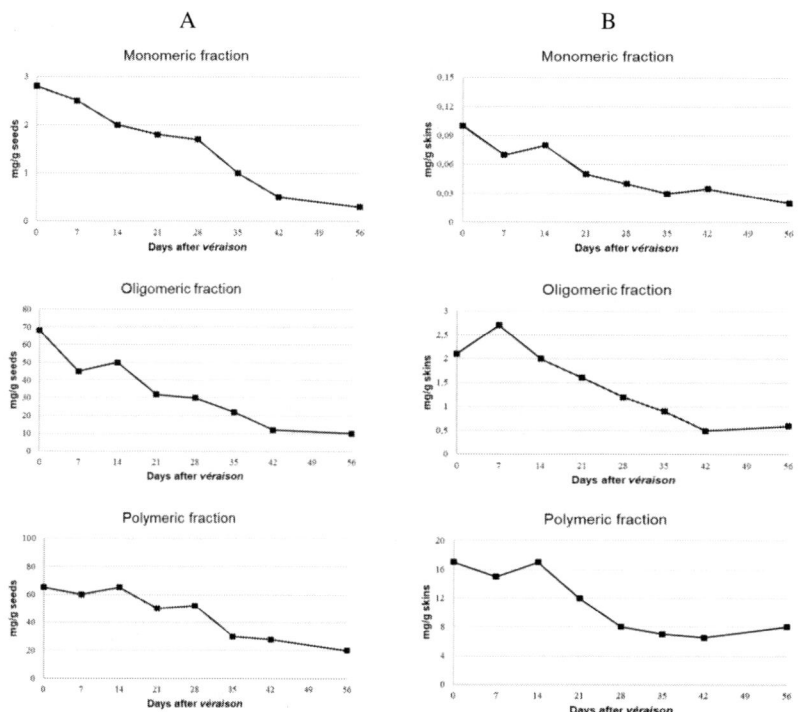

Figure 5. Evolution of the different proanthocyanidin fractions of seeds (A) and skins (B) during the ripening of Touriga Nacional grape variety (Adapted from Jordão and Correia, 2012 [28]).

A more detailed study on the proanthocyanidin composition of Touriga Nacional grape variety during grape berry maturation was performed by Mateus et al. (2001a) [32]. Table 6 shows the Touriga Nacional grape berry skins and seeds (-)-catechin, (-)-epicatechin, and (-)-epicatechin gallate, procyanidin dimers, and trimer C1 levels from *véraison* to berry maturation.

Table 6. Concentrations of catechins and procyanidin oligomers (mg/g dry weight) in Touriga Nacional grape skins and grape seeds at *véraison* and during grape maturation (Harvest 1997) (Adapted from Mateus et al., 2001a [32])

	Sampling data			
	18/07	2/09	08/09	15/09
Skins				
Catechins				
(+)-catechin	0.040 - 0.096	0.018 - 0.037	0.012 - 0.022	0.012 - 0.021
(-)-epicatechin	0.020 - 0.044	tr - 0.013	tr - 0.010	Tr
Dimers				
B1	0.110 - 1.174	0.067 - 0.326	0.155-0.175	0.184 - 0.260
B2	0.034 - 0.071	0.071 - 0.033	0.019 - 0.012	tr - 0.020
B3	0.016 - 0.066	tr - 0.016	tr - 0.011	nd - 0.013
B4	nd - 0.050	nd	nd	Nd
Catechins + dimers	0.260 - 1.464	0.172 - 0.425	0.195-0.230	0.234 -0.297
Total extractable proanthocyanidins	124-193	100-125	101-109	86-113
Seeds				
Catechins				
(+)-catechin	3.63 - 5.33	0.46 - 1.52	0.67 - 0.88	0.71 - 0.80
(-)-epicatechin	6.49 - 7.93	0.75 - 2.56	0.80 - 0.92	1.39 - 1.92
(−)−epicatechin *O*-gallate	3.40 - 5.90	0.15 - 0.44	0.30 - 0.46	0.26 - 0.39
Dimers				
B1	1.42 - 1.71	2.76 - 4.00	2.56 - 3.82	2.36 - 3.58
B2	0.35 - 0.4	5.28 - 6.89	2.13 - 7.14	4.86 - 5.23
B3	1.26 - 1.90	0.63 - 0.86	0.82 - 1.0	0.56 - 0.96
B4	2.31 - 2.43	1.49 - 2.27	2.11 - 2.45	1.54 - 2.57
B5	0.63 - 0.74	078 - 0.88	0.73 - 1.06	0.64 - 1.05
B6	0.61 - 2.02	0.31 - 0.62	0.74 - 1.56	nd - 1.91
B7	nd - 0.25	nd - 0.13	nd	nd - 0.28
B8	0.07 - 0.09	0.57 - 0.76	0.57 - 071	0.51 - 0.71
B2-gallate	17.70 - 28.50	1.89 - 2.42	1.78 - 3.59	1.59 - 1.97
Trimer C1	0.64 - 0.92	1.27 - 1.80	1.39 - 1.82	1.10 - 1.43
Catechins +dimers+trimer C1	42.07 - 54.59	19.63 - 21.86	14.72- 25.2	16.33 - 21.99
Total extractable proanthocyanidins	168 - 213	100 - 154	93 - 169	110 – 131

nd = not detected; tr = traces.

It is shown that in Touriga Nacional grape skins only procyanidins linked through a C_4-C_8 bond (dimers B1 to B3) were detected, opposite in Touriga Nacional grape seed, i.e., higher diversity and concentration of low molecular weight procyanidins was observed. However, grape seeds dimers linked through the C_4-C_8 bond were also the most abundant during grape berry maturation (Table 6).

Detailed Phenolic Composition of Touriga Nacional Grape Variety

The levels of total phenols, flavonoids, non-flavonoids, and total anthocyanins of Touriga Nacional grape berries and musts are shown in Table 7. As can be observed the levels of total phenols ranged from 270 to 930 mg/kg of berry, with flavonoids considered as the main phenolic group that ranges from 210 to 750 mg/kg of berry, and non-flavonoids phenols levels ranging from 60 to 180 mg/kg of berry. Among the flavonoids phenols, anthocyanins are the main phenolic compounds with levels between 350 and 760 mg/kg of berry. Anthocyanins are exclusively present in the cell walls and vacuoles of grape skins and they are directly responsible for the grapes and young red wine color. Indirectly, as a result of the formation of complexes with other polyphenols by copigmentation, they are responsible for the final color of matured wines [33].

Table 8 shows the composition of monomeric anthocyanins present in Touriga Nacional and the range of the levels found. As expected, the main anthocyanins are the 3-*O*-glucosides of malvidin, followed by the acetylglucosides, and finally by the coumaroylglucosides.

Touriga Nacional also contains considerable levels of procyanidins (Table 9), as expected mainly present in the seeds and skins with low levels in the pulp (total procyanidins from 100 to 370 mg/kg). The levels of monomeric flavonols in the seeds ranged from 300-1100 mg/kg, and in the skins from 100-200 mg/kg. Oligomeric procyanidins were composed of dimer B1 to B8 and the level of oligomeric flavanols ranges from 3300 to 7700 mg/kg in the seeds and from 100 to 2100 mg/kg in the Touriga

Nacional skins. The seeds presented higher levels of polymeric flavanols (21800 to 27100 mg/kg) when compared with the skins (2360 to 3300 mg/kg) with an mDP of 3.8-16.3 and 7.5-31.4, respectively. Also, the % of galloylation of the seed procyanidins was higher than the observed for the skin procyanidins and ranged from 16.4 to 40.1 in grape seeds, and from 3.6 to 7.2 in grape skins. On the other hand, as expected, only skins contained a considerable amount of prodelphinidin containing proanthocyanins, ranging from 6.1 to 28.5%.

Polyphenol accumulation is explained not only by the type of grape variety but also related to the grapes growing conditions, among them environmental factors, such as climate, which largely influenced the grape phenolic composition [37, 38]. Therefore, some works had studied the altitude effect, as temperature and humidity are strictly related to vineyard altitude, on Touriga Nacional grape production, and the grape phenolic composition (anthocyanin and procyanidins). Low growing altitudes (100 to 150 m) favored the accumulation of catechin monomers ((+)-catechin, (-)-epicatechin gallate), procyanidin dimers, trimer C1, as well as total extractable proanthocyanidins in Touriga Nacional grape skin when compared to grapes grown in higher altitudes (250 to 350 m) [32, 35], (Table 10). Higher growing altitudes are related to lower temperatures and higher humidity, and these conditions affect grape berry maturation, reducing grape berry polyphenolic compounds composition [32, 35].

In opposite, for the grape skin anthocyanins content, it was observed that at higher vineyard altitudes appeared to be advantageous, resulting in higher concentrations of anthocyanidin-3-O-glucosides in grapes [39] (Table 11). The observed differences in Touriga Nacional red grape composition concerning the procyanidins and anthocyanins grown at different altitudes can be explained by the fact that higher altitudes present lower temperatures and higher humidity, which affects grape maturation, decreasing the phenolic compounds content [18, 32, 35, 39].

Table 7. Touriga Nacional grape berry and must phenolic compounds

Touriga Nacional	Portugal			Brazil
	Dão Jordão and Correia (2012) [29]	Dão Costa et al. (2015) [12]	Douro Costa et al. (2015) [12]	Vale de S. Francisco Oliveira et al. (2018) [14]
Total phenolic compounds (mg/kg berry)	450	790	270	600 - 930
Flavonoids (mg/kg berry)		710	210	320 - 750
Non-flavonoids (mg/kg berry)		80	60	110 - 180
Total anthocyanins (mg/kg berry)		590	760	350 - 460
Color intensity (u.a)[1] (grape must)				6.38 - 16.16
Hue (grape must)				0.494 - 0.629

[1] a.u. - absorbance units

Table 8. Levels of monomeric anthocyanins from Touriga Nacional grape variety (from different regions and harvests)

Monomeric anthocyanins	Portugal			Brazil
	Silva and Queiroz (2016) [34] Harvest, 2012 Dão	Costa et al. (2015) [11] Harvest, 2011 Dão	Costa et al. (2015) [11] Harvest, 2011 Douro	Oliveira et al. (2018) [13] Harvest, 2016 Vale de S. Francisco
Non-acylated	mg/kg liophilized grapes	mg/g of berry	mg/g of berry	mg/g
Delphinidin 3-O-glucoside	609.1 ± 4.0	0.034 ± 0.000	0.022 ± 0.000	23 ± 0.5
Cyanidin 3-O-glucoside	75.9 ± 0.1	0.004 ± 0.000	0.002 ± 0.000	4.1 ± 0.5
Peonidin 3-O-glucoside	427.1 ± 7.9	0.067 ± 0.001	0.043 ± 0.001	15.3 ± 0.5
Petunidin 3-O-glucoside	644.8 ± 2.5	0.056 ± 0.001	0.028 ± 0.001	20.7 ± 0.5
Malvidin 3-O-glucoside	2800.2 ± 11.2	0.311 ± 0.013	0.370 ± 0.033	40.1 ± 0.5
Acetylated				
Peonidin 3-O-acetylglucoside	…			0.8 ± 0.0
Petunidin 3-O-acetylglucoside	…	0.005 ± 0.001	0.001 ± 0.000	4.6 ± 0.1
Cyanidin 3-O-acetylglucoside	…	0.002 ± 0.000		0.3 ± 0.0
Delphinidin 3-O-acetylglucoside	…			2.9 ± 0.1
Malvidin 3-O-acetylglucoside	…	0.013 ± 0.004	0.066 ± 0.001	12.2 ± 0.1
Coumaroylated				
Peonidin 3-O-coumaroylglucoside	114.5 ± 0.3	0.006 ± 0.001	0.004 ± 0.000	0.7 ± 0.0
Petunidin 3-O-coumaroylglucoside	91.1 ± 0.9	0.004 ± 0.000	0.006 ± 0.001	0.5 ± 0.0
Delphinidin 3-O-coumaroylglucoside				8.1 ± 0.5
Malvidin 3- O-coumaroylglucoside	573.6 ± 0.7	0.065 ± 0.006	0.041 ± 0.004	5.9 ± 0.5
Total monomeric anthocyanins	5336.6	0.567	0.583	139.2 ± 1.0

Table 9. Touriga Nacional seeds, skins and pulps individual proanthocyanidins content and structural characteristics

Touriga Nacional	Portugal		Brazil
	Mateus et al. (2001b) [35]	Cosme et al. (2009) [36]	Oliveira et al. (2018) [13]
Seeds	µg/berry	mg/kg	mg/kg
(+)-catechin	24.5 - 30.4		398 - 1038
(-) epicatechin	52.8 - 66.3		430 - 702
(-) epicatechin 3-O-gallate	0.009 - 14.9		5.4 - 15.3
Procyanidin B1	81.5 - 136.0		5.7 - 71.4
Procyanidin B2	167.8 - 198.8		219 - 430
Procyanidin B3	19.3 - 36.5		5.7 - 64.6
Procyanidin B4	53.2 - 97.7		6.5 - 19.7
Procyanidin B5	22.1 - 39.9		
Procyanidin B6	nd - 72.6		
Procyanidin B7	nd - 9.7		
Procyanidin B8	17.6 - 27.0		
B1 3-O-gallate			26.3 - 82.7
B2 3-O-gallate	54.9 - 74.9		7.1 - 12.3
B2 3'-O-gallate			22.7 - 40.2
Trimer C1	38.0 - 54.3		6.1 - 15.9
Trimer 2			3.1 - 13.5
Monomeric flavanols		300	500 - 1100
Oligomeric flavanols		7700	3300 - 5200
Polymeric flavanols		27100	21800 - 23500
Total tannin mDP			8.7 - 16.3
Oligomeric flavanols mDP		3.8	
Polymeric flavanols mDP		6.2	
(%) galloylation		16.4	28.3 - 40.1
Oligomers+catechins	563.7 - 835.0		
Total proanthocyanidins	3.8 - 5.0*		25600-30100
Skins	µg/berry	mg/kg	mg/kg
(+)-catechin	0.8 - 1.1		37.6 - 72.8
(-) epicatechin	Tr		35.5 - 117.4
(-) epicatechin 3-O-gallate	Nd		0.4 - 1.6
Procyanidin B1	12.8 - 14.2		2.0 - 14.0
Procyanidin B2	nd-1.4-		26.8 - 74.2
Procyanidin B3	nd-0.9		2.5 - 4.2
Procyanidin B4	Nd		1.9 - 4.3
B1 3-O-gallate	Nd		0.5 - 4.6
B2 3-O-gallate	Nd		3.0 - 4.1
B2 3'-O-gallate	Nd		22.3 - 5.2
Trimer C1	Nd		1.9 - 8.6
Trimer 2			1.7 - 7.7
Monomeric flavanols		200	100 - 200
Oligomeric flavanols		100	1300 - 2100
Polymeric flavanols		2360	2900 - 3300
Total tannin mDP			22.1 - 31.4
Oligomeric flavanols mDP		7.5	

Touriga Nacional	Portugal		Brazil
	Mateus et al. (2001b) [35]	Cosme et al. (2009) [36]	Oliveira et al. (2018) [13]
Polymeric flavanols mDP		26.4	
(%) galloylation		3.6	3.0 - 7.2
(%) prodelphinidins		28.5	6.1 - 9.4
Oligomers+catechins	14.9 - 15.3		
Total proanthocyanidins	5.97 - 6.20*		4600 - 5300
Pulps			mg/kg
Monomeric flavanols			10 - 20
Oligomeric flavanols			50 - 90
Polymeric flavanols			110 - 270
Total tannins			100 - 370
Total proanthocyanidins			100 - 370

mDP – mean degree of polymerization: * total proanthocyanidins expressed in mg/berry; nd = not detected; tr = traces.

Table 10. The concentration of catechins, procyanidin oligomers and total proanthocyanidins (µg/berry) in grape skins and seeds of Touriga Nacional grape variety at harvest at different altitudes (Adapted from Mateus et al., 2001a [32]; Mateus et al., 2001b [35])

	Low altitude (100-150 m)		High altitude (300-350 m)	
	skins	seeds	skins	seeds
Catechins				
(+)-catechin	1.1	30.4	0.8	24.5
(-)-epicatechin	tr	52.8	tr	66.3
(-)-epicatechin-O-gallate	nd	14.9	nd	0.009
Dimers				
B1	14.2	136.0	12.8	81.5
B2	nd	198.8	1.4	167.8
B3	nd	36.5	0.9	19.3
B4	nd	97.7	nd	53.2
B5	nd	39.9	nd	22.1
B6	nd	72.6	nd	nd
B7	nd	Nd	nd	9.7
B8	nd	27.0	nd	17.6
B2 O-gallate	nd	74.6	nd	54.6
Trimer C1	nd	54.3	nd	38.0
Oligomers+catechins	15.3	835.0	14.9	563.7
Total proanthocyanidins (mg/berry)	6.20	5.0	5.97	3.80

nd = not detected; tr = traces (<0.7).

Detailed Aroma Composition of Touriga Nacional Grape Variety

The wine aroma is the result of some hundreds of volatile compounds, and their concentration could range from 10^{-10} to 10^{-1} g/kg [40, 41]. Alcohols, esters, acids, aldehydes, phenols, ketones, lactones, and terpenoids are the most common volatile compounds identified in grapes, actively contributing to the wine olfactive character [42].

Table 11. Anthocyanins concentration (mg per berry) in the Touriga Nacional grape skins grown at low and high altitudes at harvest date from three vintages (1997, 1998 and 1999) (Adapted from Mateus et al. 2002 [39])

Anthocyanins	Low altitude (100-150 m)	High altitude (300-350 m)
Dp-3-*O*-glc	0.06 - 0.14	0.11 - 0.22
Cy-3-*O*-glc	0.01 - 0.06	0.02 - 0.09
Pt-3-*O*-glc	0.06 - 0.12	0.11 - 0.21
Pn-3-*O*-glc	0.07 - 0.17	0.12 - 0.19
Mv-3-*O*-glc	0.38 - 0.59	0.56 - 1.05
Dp-3-*O*-acetylglc	0.01 - 0.05	0.02 - 0.08
Cy-3-*O*-acetylglc	nd - 0.05	nd - 0.06
Pt-3-*O*-acetylglc	0.02 - 0.06	0.02 - 0.08
Pn-3-*O*-acetylglc	0.01 - 0.06	0.02 - 0.07
Mv-3-*O*-acetylglc	0.09 - 0.14	0.12 - 0.21
Dp-3-*O*-coumaroylglc	nd - 0.05	nd - 0.06
Cy-3-*O*-coumaroylglc	0.01 - 0.05	0.01 - 0.06
Pt-3-*O*-coumaroylglc	0.01 - 0.06	0.03 - 0.06
Pn-3-*O*-coumaroylglc	0.02 - 0.08	0.04 - 0.08
Mv-3-*O*-coumaroylglc	0.07 - 0.26	0.13 - 0.22
Mv-3-*O*-caffeoylglc	0.01 - 0.06	0.03 - 0.07
Total anthocyanins	1.25 - 1.57	1.40 - 2.30

Dp, delphinidin; Cy, cyanidin; Pt, petunidin; Pn, peonidin; Mv, malvidin; glc, glucoside; nd = not detected.

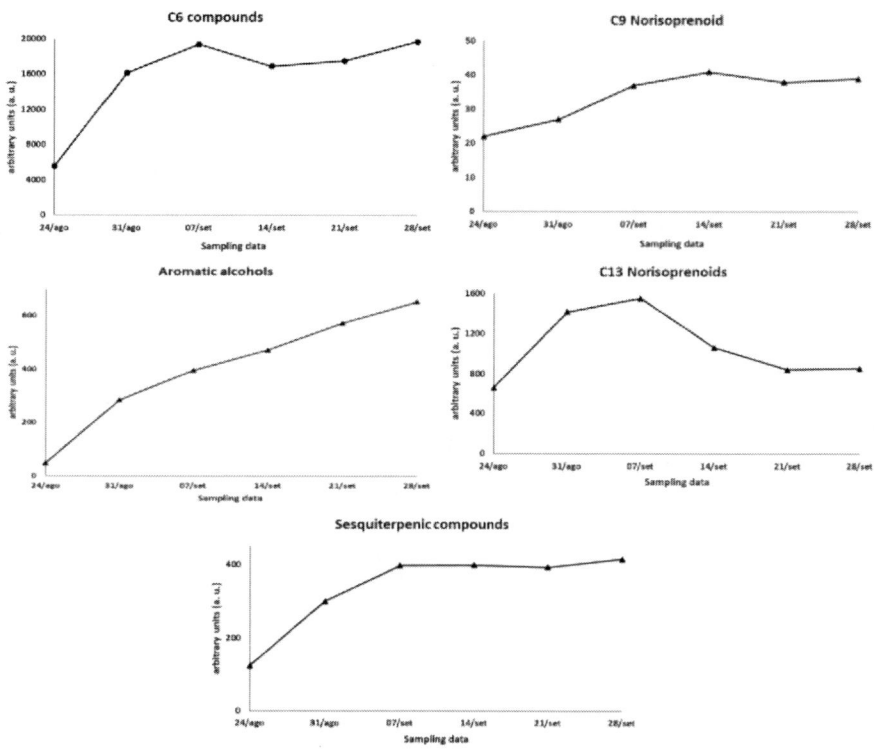

Figure 6. Evolution of the different volatile compounds from the six chemical families during the Touriga Nacional grape variety maturation (Adapted from Petronilho, 2015 [12]).

Most of the compounds associated with the varietal aroma are present as odorless precursors in grapes. Monoterpenes, norisoprenoids, benzenoids, and lactones can be present in grapes as glycosides or precursors [43]. The floral aroma of Moscatel, for example, is associated with the high levels of monoterpenes, namely, linalool, geraniol, nerol, and piranic oxides of linalool [44, 45]. Another important family of wine aroma compounds is the norisoprenoids compounds resulting from carotenoids by enzymatic and chemical degradation (photo-oxidation, oxidation, or thermal degradation).

The C13 norisoprenoids, like β-damascenone, vitispirane, α-ionone, and β-ionone contribute to the enhancement of the wine aroma complexity providing tea, lime, honey, ananas, and violet notes [46, 47]. Some grape varieties have some typical aroma compounds, such as Cabernet Sauvignon and Sauvignon Blanc, with herbaceous and green pepper character associated with the presence of 2-metoxi-3-isobutilpyrazin [48].

The volatile compounds determined in Touriga Nacional grape variety during maturation are C6 compounds (hexanal, 3-hexenal, 2-hexenal, 3-hexen-1-ol, 2-hexen-1-ol, 1-hexanol and 2,4-hexadienal), aromatic alcohols (benzyl alcohol, α,α-dimethyl benzyl alcohol, 2-phenylethanol), C9 norisoprenoid (norinone), monoterpenic compounds (α-pinene, dehydroxylinalooloxide, β-pinene, 3-carene, *m*-cymene, limonene, 1,8-cineole, β-ocimene, linalool oxide (isomer), dihydromyrcenol, α-terpinolene, linalool oxide (isomer), dihydrolinalool, linalool, rose oxide (isomer), hotrienol, β-terpineol, nerol oxide, borneol, menthol, terpinen-4-ol, *p*-cymen-8-ol, α-terpineol, verbenone, *p*-menth-1-en-9-al, geraniol (isomer), β-citronellol, citral (isomer), carvone, citral (isomer)), C13 norisoprenoids (vitispirane, theaspirane (isomer), β-damascenone (isomer), geranylacetone, 5,6-epoxy-β-ionone, 3,4-dehydro-β-ionone, α-iso-methyl ionone, β-ionone, methyl dihydrojasmonate), sesquiterpenic compounds (δ-elemene, α-copaene, β-bourbonene, aromadendrene, α-farnesene, calamenene, α-calacorene, nerolidol, globulol, caryophyllene oxide, β-eudesmol), and diterpenic compound (phytol) [12]. The evolution of the 6 chemical families of volatile compounds during grape berry maturation of Touriga Nacional grape variety from the Bairrada Demarcated Region [12] is shown in Figure 6 and the evolution of the individual compounds is shown in Table 12.

Table 12. Volatile compounds expressed as arbitrary units (a. u.) from *Vitis vinifera* L. cv. Touriga Nacional grape variety, during maturation in the Bairrada Demarcated Region (Adapted from Petronilho, 2015 [12])

	Sampling data					
	24/08	31/0	07/09	14/09	21/09	28/09
C6 compounds						
Hexanal	1501.7	3015.8	4524.9	3430.1	4203.9	4409.4
3-Hexenal	357.5	1662.9	1172	980	612.2	631.4
2-Hexenal	2061.2	3316.1	6881.8	6089.1	6860.5	4446.2
3-Hexen-1-ol	310	1125.3	771	388	122	195.2
2-Hexen-1-ol	643.5	4076.4	2854.6	1887.7	2126.9	1163
1-Hexanol	542.4	1725.7	2260.7	3426.2	3287.8	8408.4
2,4-Hexadienal	120.7	1223.9	942.3	714.5	319.6	437.5
Aromatic alcohols						
2-Phenylethanol	14.8	148.5	232.8	175.8	245.3	391
C9 Norisoprenoid						
Norinone	21.6	27.3	37.2	40.5	37.9	38.5
Monoterpenic compounds						
α-Pinene	31.5	37	53.6	109.6	118.2	7.2
Dehydroxylinalooloxide	0	29.2	35.2	0	0	0
β-Pinene	12.6	26.2	31.7	23	13.6	13.1
3-Carene	64.1	46.1	22.5	40.6	0	0
m-Cymene	58.7	70.8	44.6	23.7	14.6	12.2
Limonene	tr	94.7	31.2	29.5	31.4	73.5
1,8-Cineole	80.2	70.4	51	52	55.4	60.7
β-Ocimene	33.2	202.8	163.3	132.5	124.3	185.9
Linalool oxide (isomer)	88.1	51.7	63.7	21.1	12.1	16.9
Dihydromyrcenol	tr	296.5	314.8	231.6	265.6	226.9
α-Terpinolene	52.9	72.8	44.2	19.6	31.5	27.7
Linalool oxide (isomer)	118.1	32.7	25.8	13.4	12.4	18.7
Dihydrolinalool	tr	28.3	15.3	17.4	14.4	10.7
Linalool	460.0	573.7	628.5	792	799.1	843.7
Rose oxide (isomer)	17.8	45.9	8	8.8	0	0
Hotrienol	479.1	479.9	635.7	403	424.9	430
β-Terpineol	68.3	29.1	30.7	0	0	0
Nerol oxide	21.3	66.5	90.8	38.8	30.6	36.7
Borneol	44.0	71.2	104.9	51.4	92.4	75.5
Menthol	tr	127.4	119.9	167.1	118.9	99.6
Terpinen-4-ol	36.1	52.4	62.3	34.7	18.2	16.6
Cymen-8-ol	0	0	46.7	22.2	22.9	27
α-Terpineol	160	417.5	393.9	364.6	238.8	278.5
Verbenone	25.4	45.8	40.3	30.6	30.5	24.3
Menth-1-en-9-al	0	11.5	18.1	15.4	97.9	21.8

Table 12. (Continued)

	Sampling data					
	24/08	31/0	07/09	14/09	21/09	28/09
Geraniol (isomer 1)	tr	68.8	106.9	158.3	83.6	174.3
β-Citronellol	tr	77.0	156.9	184.6	60.3	41.2
Geraniol (isomer 2)	284.5	631.4	777.9	805.2	800.2	789.7
Citral (isomer 1)	0	21.2	31.3	41.9	50.5	51.5
Carvone	tr	47.2	48	28.5	26.4	52.1
Citral (isomer 2)	tr	57.5	58.9	47	36.6	70.5
C13 Norisoprenoids						
Vitispirane	78.7	93.7	79.9	0	0	0
β-Damascenone (isomer)	102.2	154.5	288.2	170.0	167.6	204.2
β-Damascenone (isomer)	147.4	516.1	516.1	291.7	291.2	278.0
Geranylacetone	312	492.3	488.6	508.1	294.7	272.6
5,6-Epoxy-β-ionone	tr	22.8	8.9	8.4	6.5	6.6
α-Iso-methyl ionone	16.9	19.4	14.8	0	0	0
β-Ionone	0	74	92.1	18.5	18.5	27.5
Methyl dihydrojasmonate	0	43.5	61.9	66.6	64.5	63.8
Sesquiterpenic compounds						
δ-Elemene	0	0	0	tr	11.7	10.9
α-Copaene	0	20.2	28.5	26.7	29.4	26.0
β–Bourbonene	0	17.3	23.8	24.5	28.0	31.3
Aromadendrene	17.7	16.4	17.8	23.9	20.8	38.5
α–Farnesene	0	0	8.4	10.7	13.5	19.5
Calamenene	0	tr	15.4	18.4	21.3	16.3
α-Calacorene	0	0	8.4	10.4	13.9	19.3
Nerolidol	tr	98.8	89.9	81.2	82.4	103.9
Globulol	22.2	22.0	30.2	90.0	76.3	54.7
Caryophyllene oxide	0	0	18.7	21.3	17.0	19.4
β-Eudesmol	83.2	125.2	156.1	91.4	78.8	74.8

tr-trace

Carotenoids and C13-Norisoprenoids

The level of carotenoids on the grape berries changes between grape varieties, climatic conditions, and vine management. The carotenoids are synthesized by the vine plant since the beginning of fruit growing at the *véraison* and degraded between the *véraison* and fruit maturation by oxygenases enzymes resulting in the C13 norisoprenoids compounds [49].

The lutein and β-carotene are the principal carotenoids present in red grape varieties from *Vitis vinifera* L., representing 85% of the total carotenoids, and the remaining fraction made up of xanthophylls, neochrome, neoxanthin, violaxanthin, luteoxantine, flavoxantine, zeaxantine, and lutein-5,6-oxide [49, 50].

The levels of C13-norisoprenoids precursors of Touriga Nacional are shown in Table 13 and include neoxanthin, violaxanthin, luteoxanthin, lutein, and β-carotene. The levels and relative amount of carotenoids in Touriga Nacional grape skin are affected by training systems.

Table 13. C13-Norisoprenoids precursors of Touriga Nacional grapes in the harvest day in two vine training systems (Douro Valley) (Adapted from Sousa, 2010 [51])

Training systems	Touriga Nacional		
	VSP	LYS	Corresponding C13 Norisoprenoids
Compounds	µg/kg	µg/kg	
Neoxanthin[a] 1	nd	61.1	β-damascenon
Violaxanthin[a]	12.1	18.6	vitispirane
Luteoxanthin[a]	2.4	16.1	
Lutein	515.8	423.8	
β-carotene	737.0	738.8	β-ionon, β-damascenon
Carotenoids	**1267.3**	**1258.5**	

[a] concentration expressed in lutein equivalent (µg/kg). Training systems: VSP (vertical shoot positioning), LYS (double lanyard system); nd = not detected.

Table 14. C13-Norisoprenoids in grapes of Touriga Nacional (glycosylated form) (Adapted from Sousa, 2010 [51])

Training systems	Touriga Nacional	
	VSP	LYS
Compounds	µg/kg	µg/kg
β-damascenone	1.83	2.79
TDN	4.63	3.27
TPB	1.27	0.70
β-ionone	nd	0.80
Total of norisoprenoids	**7.73**	**7.56**

TDN - 1,1,6-Trimethyl-1,2-dihydronaphthalene; TPB - 1-(2,3,6-trimentilfenil) buta-1,3-diene; Training systems: VSP (vertical shoot positioning), LYS (double lanyard system); nd = not detected.

This change in C-13 norisoprenoids precursors affected the levels of individual norisoprenoids present in the grapes obtained for the two viticultural training systems (Table 14).

Terpenoids of Touriga Nacional – Free and Glycosylated Fractions

The terpenoids are compounds accumulated in the grape during the maturation, presenting the maximum level in the over-maturation [52]. The low odor detection threshold (ODT) and high level in Moscatel wine (1 a 3 mg/L) make them very important to the floral aroma of Moscatel [53].

Table 15. Terpenoids of the free and glycosylated fractions of Touriga Nacional grapes in the harvest day in two vines training systems (Douro Valley) (Adapted from Sousa, 2010 [51])

Training systems	Touriga Nacional	
	VSP	LYS
Compounds	µg/kg	µg/kg
Free fraction		
Linalool	nd	2.3
Nerol	nd	nd
α-terpeniol	nd	nd
4-terpeniol	nd	nd
HO trienol	nd	nd
α-terpinen	8.4	4.7
α-terpinolen	19.9	7.3
Total of terpenoids	**28.3**	**14.3**
Glycosylated		
Linalool	20.9	42.7
Nerol	14.9	4.0
α-terpeniol	3.2	1.0
4-terpeniol	nd	5.8
HO trienol	nd	nd
α-terpinen	1.0	2.9
α-terpinolen	2.1	6.2
Total of terpenoids	**42.1**	**62.6**

Training systems: VSP (vertical shoot positioning), LYS (double lanyard system); nd = not detected.

A significant part of the terpenoid compounds is in the non-volatile glycosylated form (90%), however, it can be hydrolyzed to the free volatile form by enzymatic or chemical pathways during the fermentation and wine aging. The terpenoids fraction (Table 15) was almost exclusively formed by α-terpinene and α-terpinolene, while the linalool and nerol are the principal in the glycosylated form. The viticultural training system has been shown to affect the terpenoid levels present in Touriga Nacional (Table 15).

INFLUENCE OF *TERROIR* AND WINEMAKING TECHNOLOGY ON TOURIGA NACIONAL WINE COMPOSITION AND SENSORY PROFILE

Touriga Nacional Wine Phenolic Composition

The phenolic composition from Touriga Nacional wine produced in different regions from Portugal and Brazil is shown in Table 16.

Table 16. Phenolic composition of Touriga Nacional wine produced in Portugal and in Brazil

Parameters	Portugal Fernandes (2009) [8]	Brazil Oliveira et al. (2018) [13]
	Range of values	Range of values
Total phenolic compounds (mg/L)	1596 - 2230	1808 - 3187
Flavonoids (mg/L)	1419 - 2043	1628 - 1674
Non-flavonoids compounds (mg/L)	175 - 241	138 - 206
Total anthocyanins (mg/L)	347.0 - 497.9	382 - 714
Colored anthocyanins (mg/L)	20.80 - 76.20	39.5 - 126.3
Color intensity (a.u.[1])	6.4 - 12.8	9.0 - 22.8
Hue	0.65 - 0.87	0.591 - 1.208
Total pigments (a.u.)[1]	19.29 - 27.88	9.6 - 51.2
Polimeric pigments (a.u.)[1]	1.94 - 3.17	2.5 - 7.3

[1] a.u. - absorbance units

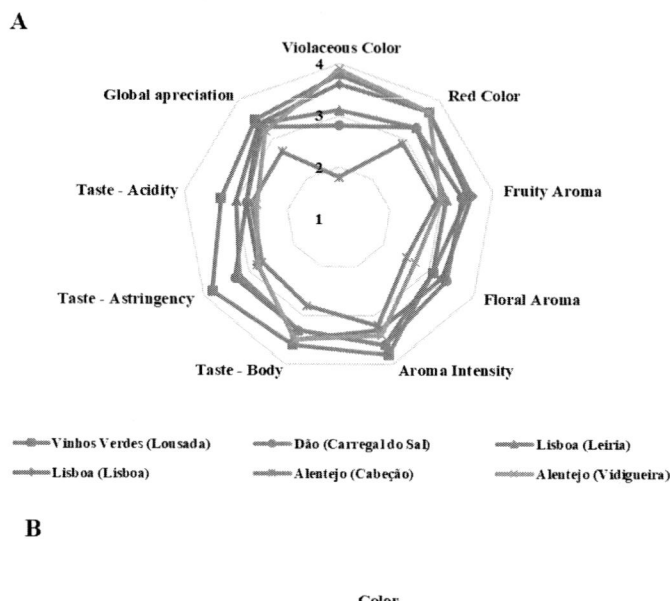

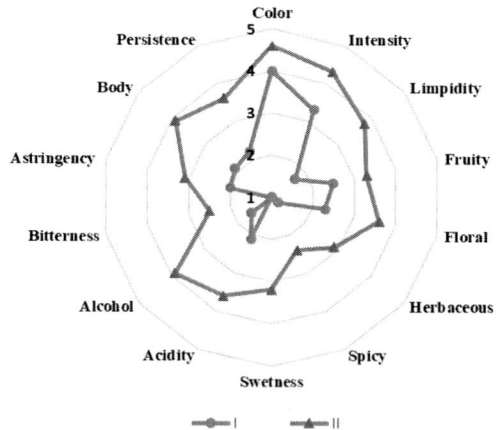

Figure 7. Sensory profile of Touriga Nacional wines produced in different regions from Portugal (Adapted from Fernandes, 2009 [8]) (A) and in semi-arid tropical, Brazil (Vale de S. Francisco). (Adapted from Oliveira et al., 2018 [13]) (B).

Figure 7 shows the sensory profile of Touriga Nacional wines produced in different regions from Portugal (Figure 7A) and Brazil (Figure 7B). In the Touriga Nacional wines produced in the different regions from Portugal, it was observed differences in the color attributes, namely in the violet wine color, with higher values in Alentejo (Vidigueira), Vinhos Verdes (Lousada), and Lisboa (Lisboa) wines, and a low value in Alentejo (Cabeção). Differences were also observed for the flower aroma, body, and astringency attributes as well as in the taste attributes [8].

The sensory profile of Touriga Nacional wines produced in the tropical and semi-arid regions of Brazil (Figure 7B) showed a higher score related to color. Regarding the aromatic attributes, wines obtained high scores for floral and fruity aroma. The floral descriptor in Touriga Nacional wines is related to the high levels of terpene compounds as well as norisoprenoids in C13 compared to other wines produced from other red grape varieties, which promotes higher floral aroma [9, 54].

The Touriga Nacional wine characteristics of six different regions, namely Vinhos Verdes (Lousada), Dão (Carregal do Sal), Lisbon (Leiria and Lisbon), and Alentejo (Cabeção and Vidigueira), of the 2008 vintage were studied by Fernandes (2009) [8]. The results presented in Tables 17, 18, and 19 are related to the wines after alcoholic and malolactic fermentations. It was observed that the content of colored anthocyanins is directly related to the wine pH, with wines from Vinhos Verdes (Lousada), and Lisboa (Lisboa), wines with a lower pH, presented a higher content of colored anthocyanins, due to higher levels of anthocyanin in the flavilium cation form (Table 17).

Regarding the polymeric pigments, the highest values are observed in the wines from Alentejo (Vidigueira), Vinhos Verdes (Lousada), and Lisboa (Lisboa). Polymeric pigments are also essential for the wine color, but they are less sensitive to the wine pH values (Tables 17 and 18).

For color intensity and hue, it was observed a similarity among Vinhos Verdes (Lousada), and Lisboa (Lisboa) wines, indicating values of wines with a younger color, also similar to the values of the Dão (Carregal do Sal), Lisboa (Leiria) and Alentejo (Cabeção) wines, which hue values showed some browning (Table 17). Concerning the values of copigmentation, they do not have significant differences from each wine. Regarding the content of non-flavonoid compounds, it is observed that the highest values are recorded in the Dão wines, and the lowest in the Alentejo (Cabeção) region (Table 19) [8].

Table 17. Data from Touriga Nacional wines after alcoholic and malolactic fermentation, alcohol content, pH, total acidity and quantification of phenolic compounds. (Adapted from Fernandes, 2009 [8])

	Alcoholic content (%v/v)	Total acidity (g tartaric acid/L)	pH	Total anthocyanins (mg/L)	Colored anthocyanins (mg/L)	Ionized anthocyanins (%)	Color intensity (a.u)[1]	Hue
Vinhos Verdes (Lousada)	14.5	8.3	3.47	432.7	76.20	17.6	12.8	0.65
Dão (Carregal do Sal)	15.3	7.1	3.57	480.3	40.00	8.3	10.4	0.87
Lisboa (Leiria)	13.5	6.7	3.69	467.1	37.20	8.0	8.4	0.80
Lisboa (Lisboa)	14.6	6.4	3.64	497.9	51.80	10.4	10.8	0.68
Alentejo (Cabeção)	13.6	5.9	4.06	347.0	20.80	6.0	6.4	0.86
Alentejo (Vidigueira)	14.0	6.0	3.81	475.9	47.00	9.9	11.3	0.74

[1] a.u. - absorbance units

Table 18. Data from Touriga Nacional wines after alcoholic and malolactic fermentation, for total pigments, polymeric pigments, polimerized pigments, color due to copigmentation (CCP) and tannin (Adapted from Fernandes, 2009 [8])

	Total pigments (a.u.)[1]	Polymeric pigments (a.u.)[1]	Polymerized pigments (%)	Color due to copigmentation (a.u.)[1]	Color due to copigmentation (%)	CCP mg malvidin-3-glucoside/L	BSA index (NTU/mL wine)
Vinhos Verdes (Lousada)	24.64	3.01	12.2	0.24	24.2	4.9	582.5
Dão (Carregal do Sal)	26.77	2.75	10.3	0.23	23.0	4.7	434.0
Lisboa (Leiria)	25.55	2.20	8.6	0.31	30.6	6.2	507.0
Lisboa (Lisboa)	27.88	2.98	10.7	0.26	26.1	5.3	461.9
Alentejo (Cabeção)	19.29	1.94	10.1	0.23	23.1	4.7	487.6
Alentejo (Vidigueira)	26.97	3.17	11.8	0.25	25.1	5.1	487.4

[1] a.u. - absorbance units

Regarding the BSA index, Table 18 shows high values for all wines, the Vinhos Verdes (Lousada), from north of Portugal, presented the highest value, while the lowest values were from the Dão (Carregal do Sal) wines, which probably indicate a more elevated wine "softness", considering that the tannin power evaluates the sensation of astringency in wines [8].

Table 19 showed the concentration of total phenolic compounds, non-flavonoids, and flavonoids compounds from the Touriga Nacional wines produced in the different Portuguese regions studied by Fernandes (2009) [8].

Table 19. Data from Touriga Nacional wines after alcoholic and malolactic fermentation, for total phenolic compounds, non-flavonoids and flavonoids (Adapted from Fernandes, 2009 [8])

	Total phenolic compounds (mg/L)	Non-flavonoids compounds (mg/L)	Flavonoids (mg/L)
Vinhos Verdes (Lousada)	2069	197	1871
Dão (Carregal do Sal)	2230	187	2043
Lisboa (Leiria)	1709	195	1514
Lisboa (Lisboa)	1865	241	1623
Alentejo (Cabeção)	1596	177	1419
Alentejo (Vidigueira)	1939	175	1764

Table 20 shows the phenolic compounds (flavonoids and phenolic acids) in commercial Touriga Nacional red wines from Brazil [55], and Touriga Nacional red wine produced in different vintages and harvests also in Brazil [13].

The results of fourteen monomeric anthocyanins determined in Touriga Nacional wines produced in Portugal and Brazil are shown in Table 21. The concentrations of monomeric anthocyanins in Touriga Nacional wines produced in Brazil [13] were lower (64.3-113.1 mg/L total monomeric anthocyanins) than those obtained by Mateus et al. (2001) [35] also in Touriga Nacional wines produced in Portugal in Douro Demarcated region (185.94-410.64 mg/L total monomeric anthocyanins).

Table 20. Phenolic compounds in Touriga Nacional red wines produced in Brazil

Compounds	Burin et al. (2011) [55]	Oliveira et al. (2018) [13]
Catechin (mg/L)	22.34 ± 0.29	4.7-13.2
Caffeic acid (mg/L)	19.88 ± 0.81	
p-Coumaric acid (mg/L)	5.08 ± 0.12	
Ferulic acid (mg/L)	1.99 ± 0.06	
Quercetin (mg/L)	9.06 ± 0.21	
Total phenols (mg/L)		1808-3187
Flavonoids (mg/L)		1628-1674
Non-flavonoids (mg/L)		138-206
Total anthocyanins (mg/L malvidin)		382-714
Colored anthocyanins (mg/L malvidin)		39.5-126.3
Ionization index (%)		8.7-25.8
Total pigments (u.a.)[1]		19.6-51.2
Polymerized pigment (u.a.)[1]		2.5-7.3
Polymerization index (%)		8.2-27.7
Copigmentation (%)		9.8-30.6
Color Intensity (u.a.)[1]		9.04-22.75
Hue		0.591-1.208
Monomeric (mg/L)		10.0-31.6
Oligomeric (mg/L)		107-796
Polymeric (mg/L)		719-1240
BSA index (NTU/mL wine)		213-352
(-) epicatechin (mg/L)		5.8-18.1
(-) epicatechin 3-*O*-gallate (mg/L)		0.1-1.1
Procyanidins dimers B1 (mg/L)		1.6-12.7
Procyanidins dimers B2 (mg/L)		8.2-16.3
Procyanidins dimers B3 (mg/L)		1.0-3.8
Procyanidins dimers B4 (mg/L)		0.1-6.2
B1 3-*O*-gallate (mg/L)		0.1-1.0
B2 3-*O*-gallate (mg/L)		0.4-1.6
B2 3'-*O*-gallate (mg/L)		0.7-1.5
Trimer C1 (mg/L)		0.5-2.6
Trimer 2 (mg/L)		0.1-2.8

[1] a.u. - absorbance units

Table 21. Monomeric anthocyanins profile from wines produced from Touriga Nacional grape variety

Monomeric anthocyanins	Portugal		Brazil
	Mateus et al. (2001) [35]	Mateus et al. (2001) [35]	Oliveira et al. (2018) [13]
	Altitude 100-150 m	Altitude 300-350 m	
Non-acylated	mg/L	mg/L	mg/L
Delphinidin 3-O-glucoside	2.45	13.47	1.7 - 3.3
Cyanidin 3-O-glucoside	0.66	1.32	0.4 - 3.2
Peonidin 3-O-glucoside	8.78	26.01	0.5 - 5.3
Petunidin 3-O-glucoside	5.62	11.55	0.7 - 3.9
Malvidin 3-O-glucoside	87.33	217.41	28.2 - 78.5
Acetylated			
Peonidin 3-O-acetylglucoside	1.38	6.79	0.6 - 2.3
Petunidin 3-O-acetylglucoside	2.79	10.77	0.9 - 3.3
Cyanidin 3-O-acetylglucoside	5.28	13.23	nd - 1.2
Delphinidin 3-O-acetylglucoside	11.36	13.23	0.8 - 1.1
Malvidin 3-O-acetylglucoside	20.46	40.72	7.2 - 20.0
Coumaroylated			
Peonidin 3-O-coumaroylglucoside	2.73	7.40	nd - 0.7
Delphinidin-3-O-coumaroylglucoside	0.67	3.11	nd - 1.5
Cyanidin-3-O-coumaroylglucoside	4.00	nd	nd
Petunidin-3-O-coumaroylglucoside	0.84	2.00	0.4 - 6.0
Malvidin 3-O-coumaroylglucoside	17.70	26.95	3.9 - 5.9
Peonidin-3-O-caffeoylglucoside	2.73	7.40	nd
Malvidin-3-O-caffeoylglucoside	17.70	26.95	nd
Total monomeric anthocyanins	185.94	410.64	64.3 - 113.1

nd - not detected.

Touriga Nacional Wine Volatile Composition

Table 22 shows the volatile composition of Touriga Nacional wines from the Bairrada region [12]. A total of 71 volatile compounds, distributed over 9 chemical families, including esters, alcohols, acids, carbonyl compounds, terpenic compounds, C13 norisoprenoids, lactones, phenols, and thiols are detected in this wine (Table 22). The flowery aromas of Touriga Nacional are primarily related to the C13 norisoprenoids (β-damascenone and β-ionone). Among the monoterpenic compounds, linalool and geraniol are present in Touriga Nacional wine at high concentrations. The wines from Touriga Nacional were considered by Guedes de Pinho et al. (2007) [10] as wines with a high concentration in terpenol compounds.

The most abundant volatile compounds are esters, alcohols, and acids (mainly C_4–C_{10} fatty acids). These volatile compounds are produced during alcoholic fermentation, and they are essential for the fermentative wine's aroma [56]. From the 9 chemical families identified, esters represent the main family compound in Touriga Nacional wine (a total of 19 volatile compounds). Several authors have described these compounds as essential aroma compounds of young wine, related to their fruity and sweet attributes [57, 58].

Alcohols and acids are quantitatively the largest groups of volatile compounds present in Touriga Nacional wine (Table 22). Among these, the higher concentration is isoamyl alcohol, phenylethanol, isobutanol, and acetic acid. According to Lambrechts and Pretorius (2000) [56], isoamyl alcohol (40-70% of the total alcohol fraction) is the main aliphatic alcohol synthesized by yeast during alcoholic fermentation. The alcohols such as 1-hexanol, (Z)-3-hexen-1-ol, 1-butanol, and benzyl alcohol are at concentrations under their odor threshold values, the two first alcohols are described by Ferreira et al. (1995) [59] as harming on wine aroma quality.

The changes observed in norisoprenoids in grapes concerning the vine training systems affect the levels of norisoprenoids in wine (Table 23), especially in the content of β-ionone that is responsible for the violet aroma. Due to its low limit of perception, it is a crucial aroma of the Touriga Nacional aroma profile. It is also possible to verify that, regardless of the vine training system, the glycosylated fraction is present in a more elevated amount than the free fraction. β-ionone was the compound found in higher concentrations. Also, Guedes de Pinho et al. (2007) [10] described high levels of β-ionone when compared to the other compounds. β-ionone presents a low perception limit (90 ng/L), therefore it is one of the main compounds contributing to the aroma of Touriga Nacional wines.

Table 22. Quantitative Touriga Nacional wine volatile compounds obtained in wines produced in the Portuguese Bairrada Demarcated Region (Adapted from Petronilho, 2015 [12])

Compounds	Concentration (µg/L)	Odor threshold (µg/L)**	Descriptor
Esters			
Ethyl acetate [a]	61364.1 - 64360.2	12264	fruit
Ethyl propanoate [a]	238.9 - 242.3	10	ananas
Ethyl butyrate [a]	270.5 - 295.2	20	apple
Isoamyl acetate [a]	361.9 - 384.6	30	banana
Ethyl hexanoate [b]	622.4 - 683.9	14	green apple
Ethyl lactate [c]	72395.1 - 77143.7	154636	buttery
Ethyl octanoate [d]	288.3 - 293.4	5	sweet
Ethyl decanoate [d]	53.3 - 55.1	200	flowery
Diethyl succinate [e]	3042.0 - 3181.5	200000	light fruity
Phenylethyl acetate [d]	147.7 - 149.4	250	roses
Ethyl isobutyrate [e]	39.9 - 42.0	15	fruity
Isobutyl acetate [e]	119.7 - 125.7	1600	banana
Butyl acetate [e]	7.3 - 7.6	1880	banana
Ethyl 2-methylbutyrate [e]	5.1 - 5.3	18	fruit
Ethyl isovalerate [e]	5.0 - 5.3	3	apple
Ethyl furoate [e]	1.3 - 1.3	16000	
Ethyl dihydrocinnamate [e]	0.7 - 0.7	1.6	rock-rose
Methyl vanillate [e]	55.3 - 57.8	3000	floral
Ethyl vanillate [e]	1174.7 - 1241.1	990	floral
Carbonyl compounds			
Acetaldehyde [f]	542.0 - 556.7	500	green apple
Diacetyl [f]	297.6 - 324.9	100	butter
Acetoin [f]	193.3 - 199.3	150000	
Phenylacetaldehyde [e]	14.0 - 15.3	1	
Alcohols			
Isobutanol [f]	63013.2 - 66004.1	40000	poignant
1-Butanol [f]	1384.8 - 1571.1	150000	medicinal
Isoamyl alcohol [f]	255224.0 - 266522.1	30000	poignant
1-Hexanol [a]	1356.7 - 1432.73	8000	herbaceous
(Z)-3-Hexenol [a]	34.6 - 34.8	400	herbaceous
Benzyl alcohol [e]	56.9 - 57.5	200000	urine
Phenylethanol [e]	53554.3 - 56785.9	14000	rose
Terpenic compounds			
Linalool [e]	29.2 - 30.7	25	floral
Linalool acetate [e]	0.4 - 0.4	unknown	floral
α-Terpineol [e]	14.8 - 15.3	250	floral
β–Citronelol [e]	12.5 - 13.5	100	citrus
Geraniol [e]	39.4 - 41.5	20	floral

Table 22. (Continued)

Compounds	Concentration (µg/L)	Odor threshold (µg/L)**	Descriptor
Lactones			
γ-Butyrolactone [c]	8203.2 - 8537.7	35000	burnt plastic
(E)-Whiskylactone [e]	1.4 - 1.5	790	woody
γ-Decalactone [e]	701.7 - 704.7	88	peach
δ-Decalactone [e]	55.8 - 58.8	386	
γ-Nonalactone [e]	25.1 - 27.2	30	sweet
Acids			
Acetic acid [c]	734655.1 - 798360.9	200000	acetic acid
Isobutyric acid [c]	1975.4 - 2012.8	230	cheese
Butyric acid [c]	1386.2 - 1464.1	173	cheese
Isovaleric acid [d]	1150.4 - 1198.5	33.4	cheese
Hexanoic acid [d]	2594.9 - 2678.7	420	
Octanoic acid [d]	1865.1 - 1975.3-	500	rancid
Decanoic acid [d]	408.8 - 424.9	1000	rancid
Norisoprenoids			
β-Damascenone [e]	2.9 - 3.0	0.05	tea
β-Ionone [e]	0.5 - 0.5	0.09	violet
Volatile Phenols			
Guaiacol [e]	5.8 - 6.1	9.5	balsamic
Eugenol [e]	3.4 - 3.5	6	spicy
o-Cresol [e]	3.7 - 3.8	3	smoky
m-Cresol [e]	1.7 - 1.8	68	
4-Ethylguaiacol [e]	6.9 - 7.2	33	leather
4-Ethylphenol [e]	3.7 - 4.3	440	horse sweet
4-Vinylguaiacol [e]	11.5 - 12.7	1100	mint
2,6-Dimethoxyphenol [e]	19.7 - 20.0	-	smoky
4-Vinylphenol [e]	1.2 - 1.5	180	pharmaceutic
4-Allyl-2,6-dimethoxyphenol [e]	10.1 - 10.6	120	bacon
Acetovanillone [e]	266.8 - 278.0	1000	vanilla
Thiols			
Methionol [e]	2629.5 - 2749.1	1000	raw-potato
2-Methyl-3-furanthiol [g]	0.347 - 0.408	0.0050	cooked meat
2-Furfurylthiol [g]	0.002 - 0.002	0.0004	boiled meat
4-Mercapto-4-methyl-2-pentanone [h]	0.010 - 0.012	0.0008	roasted meat
3-Mercaptohexyl acetate [h]	0.001 - 0.002	0.0042	tropical fruity
3-Mercapto-1-hexanol [g]	0.022 - 0.047	0.0600	tropical fruity
Benzylmercaptan [g]	0.003 - 0.004	0.0003	

* concentration of wines volatile components was obtained by dividing the chromatographic area of each volatile component by the area of the corresponding internal standard: (a) 4-methyl-2-pentanol; (b) ethyl heptanoate; (c) 4-hydroxy-4-methyl-2-pentanone; (d) heptanoic acid; (e) 2-octanol; (f) 2-butanol; (g) 4-methoxi-α-toluenothiol; (h) 1,4-dithioerythritol octafluoronaphthalene (OFN). ** a Odor threshold values previously reported in the literature (for mixtures of ethanol/water; Campo et al. (2006) [60] and Gómez-Míguez et al. (2007) [58]. The odor descriptors are from Costa (2004) [61] and PubChem (2020) [63].

Another study confirmed the importance of the norisoprenoids compounds on the Touriga Nacional wine aroma, especially β-damascenone, which presented the higher OSV (olfactory spectrum value) in Touriga Nacional wine aroma analysis by the charm methodology [61]. As previously mentioned the level of this compound on the grapes is affected by sunlight exposure and depends on the climate conditions and vine practices. In this context, the exposition to light at the *véraison* is favorable to carotenoids synthesis, and after the *véraison*, it affects the nature of norisoprenoids produced [63].

Table 23. Free and glycosylated fractions of C13-Norisoprenoids in Port wine produced with Touriga Nacional (Adapted from Sousa, 2010 [51])

Training systems	Touriga Nacional	
	VSP	LYS
Compounds	µg/L	µg/L
Free fraction		
β-damascenone	0.04	nd
TDN	nd	nd
TBP	nd	nd
β-ionone	0.01	0.44
Total of norisoprenoids	**0.05**	**0.44**
Glycosylated fraction		
β-damascenone	0.06	0.59
TDN	0.44	0.50
TBP	0.02	0.11
β-ionone	11	0.63
Total of norisoprenoids	**0.63**	**1.83**

TDN - 1,1,6-Trimethyl-1,2-dihydronaphthalene; TPB - 1-(2,3,6-trimentilfenil) buta-1,3-diene; Training systems: VSP (vertical shoot positioning), LYS (double lanyard system), nd - not detected.

The Touriga Nacional typical floral bergamot aroma can be significantly enhanced as a result of its precursor hydrolysis, catalyzed by enological commercial enzymes richer in glycosidases. Enzyme treatments of the fuller-bodied Touriga Nacional wines resulted in considerable enhancement of the floral-citrus bergamot aroma intensity. Analytical evaluations indicated that both linalool and geraniol are likely contributors, which partly explain the enhancement of the "earl-grey" character [64].

These results suggest that the commercial glycosidase enzymes, and the application time, as well as the stop of the enzyme activity, need to be optimized to get an increase in the quality or wine preference. Also, blending operations should help achieve the desired effect [64].

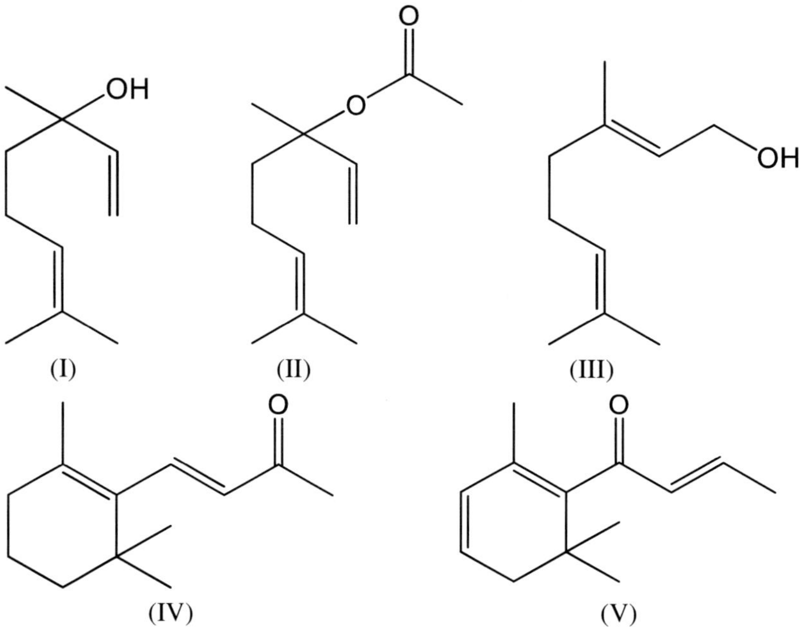

Figure 8. Structures of key-aroma compounds of Touriga Nacional: (I) linalool; (II) linalyl acetate; (III); geraniol; (IV) β-ionone; (V) β-damascenone.

After fermentation, the levels of free terpenes in Touriga Nacional wines increase significantly for both vine training systems (Table 24), with the vertical shoot positioning presenting more than 320 µg/L. Linalool, α-terpeniol, and α- terpineol are the principal compounds, with the linalool presenting more than 100 µg/L, higher than its olfactory limit perception (25 µg/L). Linalool has been shown an important component of the varietal aroma of Touriga Nacional red grape variety, contributing to the bergamot or earl-grey like aroma typicity [64].

The floral and citric aromas of Touriga Nacional grape variety were confirmed by the analyses of seventy-five Touriga Nacional red wines. It was observed that the levels of free terpenoids were higher than those from Touriga Franca, Tinta Roriz, Tinta Barroca and Tinto Cão monovarietal red wines [10], with the sensory panel experts from wine industry attributing to Touriga Nacional high-quality wines the "bergamot-like" aroma descriptor.

Table 24. Free and glycosylated terpens of Port wine produced with Touriga Nacional grape variety (Adapted from Sousa, 2010 [51])

Training systems	Touriga Nacional	
	VSP	LYS
Compounds	µg/L	µg/L
Free		
Linalool	176.8	15.1
Nerol	nd	nd
α-terpeneol	85.8	128.6
α-terpinen	42.6	0.6
α-terpinolen	12.0	nd
Total of terpens	**317.2**	**144.3**
Glycosylated		
Linalool	170.3	nd
Nerol	105.3	418.7
α-terpeneol	544.4	640.3
α-terpinen	24.3	48.1
α-terpinolen	17.3	20.9
Total of terpens	**861.6**	**1128.0**

Training systems: VSP (vertical shoot positioning), LYS (double lanyard system); nd - not detected.

Touriga Nacional Wine – Winemaking Technology

The effect of the maceration time on Touriga Nacional phenolic composition was studied with Touriga Nacional grown in São Franciso Valley (Brazil). Six sampling points during maceration (21 day at 25 °C) were studied: MS0 (time zero), MS1 (four days), MS2 (eight days), MS3 (thirteen days), MS4 (eighteen days), and MS5 (twenty-one days). The content of flavan-3-ols, stilbenes, and flavonols increased until MS3, MS2, and MS1, respectively, and did not change afterward. The concentration of monomeric anthocyanins increased until stage MS1 (four days), then remained constant until MS2 (eight days), and decreased afterward. The final wine had 48% less monomeric anthocyanins compared with MS2, the highest decrease observed for malvidin-3-glucoside. Phenolic acids showed a peak concentration at MS1 until MS2 with further decrease starting at MS3. Such changes were due to the loss of caftaric acid, the major compound at MS0, MS1, MS2, MS3, MS4, and MS5. Caftaric acid decreased from 488.33 mg/L (MS2) to 4.45 mg/L. the latter referring to wine (W). Therefore, malvidin became the major compound in the produced wine. The highest concentration of total phenolics was found at MS2 (1053.50 mg/L), and the wine (W) showed a lower concentration (616.09 mg/L). Therefore, eight days of maceration is recommended to produce young wines from Touriga Nacional since they present the highest content of phenolic compounds [65].

The cold maceration technology in Touriga Nacional seems to give some positive results concerning the color intensity (Table 25), however, with only 3 days, no statistical differences were found in the traditional process (without maceration). The sensory fruity and grape varietal aroma was also enhanced by the cold maceration in the same work [66].

Table 25. Effect of cold maceration on phenolic compounds and color in Touriga Nacional, Campana Gaúcha Region (Brazil), harvest 2017 (adapted from Langbecker, 2018 [66])

Touriga Nacional			CV %
Parameters	Traditional	Cold maceration (3 days at 8°C)	
Total anthocyaninins (mg/L)	310.62[a]	288.45[a]	11.79
Total tannins (g/L)	0.59[a]	0.72[a]	41.32
Total polyphenol index (TPI)	47.00[a]	43.76[a]	5.93
OD $A_{420\,nm}$	6.89[a]	6.95[a]	4.78
OD $A_{520\,nm}$	10.95[a]	11.58[a]	4.95
OD $A_{620\,nm}$	3.17[a]	3.27[a]	5.70
Color intensity	21.01[a]	21.80[a]	4.94
Color hue	0.629[a]	0.600[b]	1.39

Conclusion

Touriga Nacional is an outstanding Portuguese grape variety that produces original, elegant, and rich red wines, and it is starting to be used in Brazil. It is a rich grape variety due to its phenolic and aroma composition, whose composition is significantly influenced by the *"terroir"* of the different Portuguese Demarcated regions. In Alentejo Demarcated Region, a warm region, the grapes present lower berry weight and less color intensity. In contrast, in Vinhos Verdes Demarcated Region, a colder region, the grapes present higher acidity and a lower pH. In Dão Demarcated Region, for example, a region with moderate temperature, the grapes present a higher level of total anthocyanins and more color intensity. Along with the Touriga Nacional maturation, it is observed an important increase in the anthocyanins content, on the contrary, a significant proanthocyanins decrease is observed after *véraison*. The altitude significantly influences the grape phenolic composition, i.e., low growing altitude (100-150 m) appears to favor the catechin and extractable proanthocyanidins accumulation, whereas higher altitudes are favorable to grape skin anthocyanins accumulation.

Touriga Nacional is an aromatic grape variety presenting more than 70 aroma compounds in monovarietal wines. The grape variety is known for its elegant and floral aroma, and citrus, tea, violet, and bergamot notes are usually detected by professional sensory panelists and consumers. The principal compounds responsible for the Touriga Nacional wine aroma are the terpenes linalool, linalyl acetate, geraniol, and the C13-norisoprenoids, β-ionone, and β-damascenone. Eight days of maceration seems to be a reasonable time to produce a Touriga Nacional wine with a good phenolic concentration and color. Cold maceration seems also to be favorable in the phenolic extraction of Touriga Nacional grape berries.

REFERENCES

[1] IVV. (2017). *Anuário Vinho e Auardentes*. Instituto da Vinha e do Vinho. [Wine and Auardentes Yearbook]

[2] IVV. (2011). *Catálogo das castas para vinho cultivadas em Portugal*. Intituto da Vinha e do Vinho Edition 978-972-8987-21-3. [*Catalog of wine varieties grown in Portugal.*]

[3] Abade, E., Guerra, J., Pereira, C. & Sousa, M. (2007). Caracterização de Castas Cultivadas na Região Vitivinícola de Trás-os-Montes: Sub-regiões de Chaves, Planalto Mirandês e Valpaços. Mirandela, Direcção Regional de Agricultura e Pescas do Norte (DRAPN). [Characterization of Grape Varieties Cultivated in the Trás-os-Montes Wine Region: Chaves, Planalto Mirandês and Valpaços subregions.]

[4] Martins, A., Carneiro, L., Gonçalves, E., Pedroso, V., Almeida, C. & Martins, S. (2009). Perspectiva sobre a origem de castas do Dão baseadas na variedade genética intravarietal. Proceedings from the Congresso Internacional dos vinhos do Dão – Inovação e desenvolvimento - "Unbottled", *Viseu*, Portugal, *1*. [Perspective on the origin of Dão varieties based on the intravarietal genetic variety. Proceedings from the International Congress of Dão wines - Innovation and development - "Unbottled", Viseu]

[5] Lima, C. M., Fernandes, D. D. S., Pereira, G. E., Gomes, A. A., Araújo, M. C. U. & Diniz, P. H. G. D. (2020). Digital image-based tracing of geographic origin, winemaker, and grape type for red wine authentication. *Food Chemistry*, *312*, 126060.

[6] Zarrouk, O., Garcia-Tejero, I., Pinto, C., Genebra, T., Sabir, F., Prista, C., David, T. S., Loureiro-Dias, M. C. & Chave, M. M. (2016). Aquaporins isoforms in cv. Touriga Nacional grapevine under water stress and recovery – Regulation of expression in leaves and roots. *Agricultural Water Management*, *164*, 167-175

[7] Brites, J. & Pedroso, V. (2000). Castas recomendadas da Região do Dão. Direção Regional de Agricultura da Beira Litoral e Centro de

Estudos Vitivinícolas do Dão. Direção Regional de Agricultura e Pescas do Dão, Nelas, Portugal. Recovered from: https://www.drapc.gov.pt/base/documentos/castas_dao.pdf. [*Recommended grape varieties from the Dão Region. Regional Directorate of Agriculture of Beira Litoral and Dão Wine Studies Center. Regional Directorate for Agriculture and Fisheries in Dão, Nelas, Portugal.*]

[8] Fernandes, P. A. C. (2009). *Comportamento agronómico e enológico das castas Touriga Nacional e Syrah em seis regiões portuguesas* (Master Thesis). Universidade de Lisboa, Portugal. [*Agronomic and oenological behavior of Touriga Nacional and Syrah varieties in six Portuguese regions*]

[9] Oliveira, C., Silva Ferreira, A. C., Barbosa, A., Guerra, J. & Guedes de Pinho, P. (2006). Carotenoid profile in grapes related to aromatic compounds in wines from Douro region. *Journal Agricultural Food Chemistry*, *71*(1), S001-S007.

[10] Guedes de Pinho, P., Falqué, E., Castro, M. & Oliveira e Silva, H. (2007). Futher insights into the floral character of Touriga Nacional wines. *Journal of Food Science*, *72*(6), 396-401.

[11] Costa, E., Silva, J. F., Cosme, F. & Jordão, A. M. (2015). Adaptability of some French red grape varieties cultivated at two different Portuguese terroirs: Comparative analysis with two Portuguese red grape varieties using physicochemical and phenolic parameters. *Food Research International*, *78*, 302-312.

[12] Petronilho, S. L. (2015). *Sustainable viticulture in Bairrada Appellation: Vineyard and harvest year effects on grapes oenological potential* (PhD Thesis). Universidade de Aveiro, Portugal.

[13] Oliveira, J. B., Faria, D. L., Duarte, D. F., Egipto, R., Laureano, O., Castro, R., Pereira, G. E. & Ricardo-da-Silva, J. M. (2018). Effect of the harvest season on phenolic composition and oenological parameters of grapes and wines CV. Touriga Naconal (*Vitis vinifera* L.) produced under tropical semi-arid climate, semi-arid climate, in the state of Pernanbuco, Brazil. *Ciência e Técnica Vitivinícola*, *33*, 145-166.

[14] OIV. (2010). Definition of vitivinicultural "terroir". Resolution OIV/Viti 333/2010.
[15] Reis, R. M. M. & Gonçalves, M. Z. (1987). O clima de Portugal. Fascículo XXXIV. Caracterização climática da região agrícola do Alentejo. Instituto Nacional de Meteorologia e Geofísica. Lisboa. [Portugal's climate. Issue XXXIV. Climatic characterization of the agricultural region of Alentejo. National Institute of Meteorology and Geophysics.]
[16] FAO/UNESCO. (1987). FAO/UNESCO Soil Map of the World. Revised Legend with corrections. *World Soil Resources Report* 60. Rome, FAO, (Reprinted as Technichal paper 20), ISRIC, Wageningen, 1994.
[17] Kliewer, W. M. (1970). Effect of day temperature and light intensity on coloration of *Vitis vinifera* L. grapes. *Journal of the American Society for Horticultural Science*, 95, 693-697.
[18] Kliewer, W. M. & Torres, R. E. (1972). Effect of controlled day and night temperatures on grape coloration. *American Journal of Enology and Viticulture*, 23, 71-77.
[19] Kliewer, W. M. (1977). Influence of temperature, solar radiation and nitrogen on coloration and composition of Emperor grape. *American Journal of Enology and Viticulture*, 28, 96-103.
[20] Tomana, T., Utsunomiya, N. & Kataoka, I. (1979). The effect of environmental temperatures on fruit ripening on the tree. II. The effect of temperature around whole vines and clusters on coloration of Kyoho grapes. *Journal of the Japanese Society for Horticultural Science*, 48, 261-266.
[21] Soares, J. M. & Leão, P. C. S. (2009). *A vitivinicultura no semiárido brasileiro*. Brasília: Embrapa Informação Tecnológica. [*Viticulture in the Brazilian semiarid region.*]
[22] Pieri, P. & Fermaud, M. (2005). Effects of defoliation on temperature and wetness of grapevine berries. *Acta Horticulturae (ISHS)*, 689, 109-116.
[23] Tarara, J. M., Lee, J., Spayd, S. E. & Scagel, C. F. (2008). Berry temperature and solar radiation alter acylation, proportion, and

concentration of anthocyanin in merlot grapes. *American Journal of Enology and Viticulture*, 59, 235-247.

[24] Pollastrini, M., Di Stefano, V., Ferretti, M., Agati, G., Grifoni, D., Zipoli, G., Orlandini, S. & Bussotti, F. (2011). Influence of different light intensity regimes on leaf features of *Vitis vinifera* L. in ultraviolet radiation filtered condition. *Environmental and Experimental Botany*, 73, 108-115.

[25] Koyam, K., Ikeda, H., Poudel, P. R. & Goto-Yamamoto, N. (2012). Light quality affects flavonoid biosynthesis in young berries of Cabernet Sauvignon grape. *Phytochemistry*, 78, 54-64.

[26] Oliveira, M., Teles, J., Barbosa, P., Olazabal, F. & Queiroz, J. (2014). Shading of the fruit zone to reduce grape yield and quality losses caused by sunburn. *Journal International des Sciences de la Vigne et du Vin*, 48, 1-9.

[27] Oliveira, C., Ferreira, A. C., Costa, P., Guerra, J. & Guedes de Pinho, P. (2004). Effect of some viticultural parameters on the grape carotenoid profile. *Journal Agricultural Food Chemistry*, 52, 4178-4184.

[28] Jordão, A. M. & Correia, A. C. (2012). Relationship between antioxidant capacity, proanthocyanidin and anthocyanin content during grape maturation of Touriga Nacional and Tinta Roriz grape varieties. *South African Journal for Enology and Viticulture*, 33, 214-224.

[29] Kennedy, J. A., Matthews, M. A. & Waterhouse, A. L. (2000). Changes in grape seed polyphenols during fruit ripening. *Phytochemistry*, 55, 77-85.

[30] Cheynier, V., Prieur, C., Guyot, S., Rigaud, J. & Moutounet, M. (1997). The structures of tannin in grape and wines and their interactions with proteins. In T. R. Watkins (Ed.), *Wine nutritional and therapeutic benefits*, (pp. 81-93). Washington: ACS Symposium Series.

[31] Valero, E., Sánchez-Ferrer, A., Varón, R. & García-Carmona, F. (1989). Evolution of grape polyphenol oxidase activity and phenolic content during maturation and vinification. *Vitis*, 28, 85-95.

[32] Mateus, N., Marques, S., Gonçalves, A. C., Machado, J. M. & Freitas, V. (2001a). Proanthocyanidin composition of red *Vitis vinifera* varieties from the Douro Valley during ripening: influence of cultivation altitude. *American Journal of Enology and Viticulture*, 52, 115-121.

[33] Revilla, I., Pérez-Magariño, S., González-San José, M. L. & Beltrán, S. (1999). Identification of anthocyanin derivatives in grape skin extracts and red wines by liquid chromatography with diode array and mass spectrometric detection *Journal of Chromatography. A*, *847*, 83-90.

[34] Silva, L. & Queiroz, M. (2016). Bioactive compounds of red grapes from Dão region (Portugal): Evaluation of phenolic and organic profile. *Asian Pacific Journal of Tropical Biomedicine*, 6(4), 315-321.

[35] Mateus, N., Proença, S., Ribeiro, P., Machado, J. M. & Freitas, V. (2001b). Grape and wine polyphenolic composition of red *Vitis vinifera* varieties concerning vineyard altitude. *Ciência e Tecnologia de Alimentos*, *3*, 102-110. [*Food Science and Technology*]

[36] Cosme. F., Ricardo-da-Silva. J. M. & Laureano. O. (2009). Tannin profiles of *Vitis vinifera* L. cv. red grapes growing in Lisbon and from their monovarietal wines. *Food Chemistry*, *112*, 197-204.

[37] Romeyer, E. M., Macheix, J. J., Goiffon, J. R., Rominiac, C. C. & Sapis, J. C. (1983). The browning capacity of grapes. 3. Changes and importanco of hydroxicinnamic acid tartaric esters during development and maturation of the fruit. *Journal of Agricultural and Food Chemistry*, *31*, 346-349.

[38] Guilloux, M. (1981). *Evolution dos composés phénoliques de la grappe pendant la maturation du raisin. Influence dos facteurs naturels* (PhD Thesis). University of Bordeaux II, France. [*Evolution of the phenolic compounds of the bunch during the ripening of the grape. Influence of natural factors*]

[39] Mateus, N., Machado, J. M. & Freitas, V. (2002). Development changes of anthocyanins in *Vitis vinifera* grapes grown in the Douro

valley and concentration in respective wines. *Journal of the Science of Food and Agriculture*, *82*, 1689-1695.

[40] Rapp, A. (1988). Wine aroma substances from gas chromatographic analysis. In H-F. Linskens & and J.F. Jackson (Eds.), *Wine analysis*, (pp. 29-66). Berlin: Springer.

[41] Canabis, J. C., Canabis, M. T., Cheynier, V. & Teissedre, P. L. (1998). Tables de composition. In C. Flanzy (Ed.), *Oenologie: fondaments scientifiques et techiniques*, (pp. 315-336). Paris: Lavoisier Tec&Doc.

[42] Polasková, P., Herszage, J. & Ebeler, S. (2008). Wine Flavour: chemistry in a glass. *Chemical Society Reviews*, *37*, 2478-2489.

[43] Sefton, M. A., Francis, I. L. & Williams, P. J. (1994). Free and Bound Volatile Secondary Metabolites of *Vitis Vinifera* Grape cv. Sauvignon Blanc. *Journal of Food Science*, *59*(1), 142-147.

[44] Bayonove, C. & Cordonnier, R. (1970a). Recherches sur l´arome du Muscat. I- Évolution des consituints volatils au cours de la maturation du "Muscat d´Alexandrie. *Annales de Technologie Agricole*, *19*, 79-93. [Research on the aroma of Muscat. I- Evolution of volatile constituents during the maturation of "Muscat d´Alexandrie. *Annals of Agricultural Technology*]

[45] Bayonove, C. & Cordonnier, R. (1970b). Recherches sur l´arome du Muscat. II-Profils aromatiques de cépages Muscat et non Muscat. Importance du linalol, chez les Muscats. *Annales de Technologie Agricole*, *19*, 95-105. [Research on the aroma of Muscat. II-Aromatic profiles of Muscat and non-Muscat grape varieties. Importance of linalool, in Muscat. *Annals of Agricultural Technology*]

[46] Winterhalter, P. (1991). 1,1,6-Trimethyl-1,2-dihydronaphthalene (TDN) formation in wine. 1. Studies on the hydrolysis of 2,6,10, 10-tetramethyl-1-oxaspirol [4.5]dec-6-ene-2,8-diol rationalizing the origin of TDN and related C13 norisoprenoids in Resieling wine. *Journal of Agricultural and Food Chemistry*, *39*, 1825-1829.

[47] Marais, J., Van Wyk, C. & Rapp, A. (1992). Effect of storage time, temperature and region on the levels of 1, 1, 6-trimethyl-1, 2-dihydronapthalene and other volatiles, and on quality of Weisser

Riesling wines. *South African Journal for Enology and Viticulture*, *13*, 33-44.
[48] Allen, M. S., Lacey, M. J., Harris, R. L. N. & Brown, W. V. (1991). Contribution of methoxypirazines to Sauvignon Blanc wine aroma. *American Journal of Enology and Viticulture*, *42*, 109-112.
[49] Razungles, A., Babic, I., Sapis, J. C. & Bayonove, C. L. (1996). Particular behavior of epoxyxeraison and maturation of grapes. *Journal of Agricultural and Food Chemistry*, *44*, 3821-3825.
[50] Marais, J., Versini, G., van Wyk, C. J. & Rapp, A. (1991). Effect of region on free and bound monoterpene and C13-norisoprenoid concentrations in Wiesser Riesling wines. *South African Journal for Enology and Viticulture*, *13*, 71-77.
[51] Sousa, L. M. R. (2010). *Precursores aromáticos em uvas: Influência da Casta e Sistema de Condução da Vinha no Perfil Aromático de Vinhos do Porto* (Master Thesis). Universidade do Porto, Portugal. [*Aromatic precursors in grapes: Influence of Grape Variety and Conduction System of the Vine on the Aromatic Profile of Port Wines*]
[52] Wilson, B., Strauss, C. R. & Williams, P. J. (1984). Changes in free and glycosidically bound monoterpenes in developing Muscat grapes. *Journal of Agricultural and Food Chemistry*, *32*, 919-924.
[53] Terrier, A. (1972). *Les compôsés terpéniques dans l´arôme des rainsins e t des vins de certaines variétés de Vitis vinifera* (PhD Thesis). Université de Bordeaux II, France. [*Terpene compounds in the aroma of rainsins and wines of certain varieties of Vitis vinifera*]
[54] Barbosa, A., Silva Ferreira, A. C., Guedes de Pinho, P., Pessanha, M., Vieira, M., Soares Franco, J. M. & Hogg, T. (2003). Determination of monoterpenes on Portuguese wine varieties. *Proceedings from the Symposium International d' Oenologie*, Bordeaux, France, 5. Recovered from https://repositorio.ucp.pt/bitstream/10400.14/5889/1/com-inter_2003_ESB_100_Silva%20Ferreira_Ant%c3%b3nio%20C%c3%a9sar_15.pdf.

[55] Burin, V. M., Arcari, S. G., Costa, L. L. F. & Bordignon-Luiz, M. T. (2011). Determination of Some Phenolic Compounds in Red Wine by RP-HPLC: Method Development and Validation. *Journal of Chromatographic Science*, *49*, 648-651.

[56] Lambrechts, M. G. & Pretorius, I. S. (2000). Yeast and its Importance to Wine Aroma – A Review. *South African Journal for Enology and Viticulture*, *21*, 97-129.

[57] Escudero, A., Gogorza, B., Melús, M. A., Ortín, N., Cacho, J. & Ferreira, V. (2004). Characterization of the aroma of a wine from maccabeo. Key role played by compounds with low odor activity values. *Journal of Agriculture and Food Chemistry*, *52*, 3516-3524.

[58] Gómez-Míguez, M. J., Cacho, J. F., Ferreira, V., Vicario, I. M. & Heredia, F. J. (2007). Volatile components of Zalema white wines. *Food Chemistry*, *100*, 1464-1473.

[59] Ferreira, V., Fernández, P., Peña, C., Escudero, A. & Cacho, J. F. (1995). Investigation on the role played by fermentation esters in the aroma of young Spanish wines by multivariate analysis. *Journal of the Science of Food and Agriculture*, *67*, 381-392.

[60] Campo, E., Ferreira, V., Escudero, A., Marqués, J. C. & Cacho, J. (2006). Quantitative gas chromatography–olfactometry and chemical quantitative study of the aroma of four Madeira wines. *Analytica Chimica Acta*, *563*, 180-187.

[61] Costa, V. A. C. F. (2004). *Caracterização do aroma de vinhos da Vitis vinifera L. var. Touriga Nacional* (PhD Thesis). Universidade de Trás-os-Montes e Alto Douro, Portugal. [*Characterization of the aroma of wines from Vitis vinifera L. var. Touriga Nacional*]

[62] PubChem. (2020). *Open Chemistry Database* [Database]. Recovered from https://pubchem.ncbi.nlm.nih.gov/compound/.

[63] Razungles, A. J., Baumes, R. L., Dufour, C., Sznaper, C. N. & Bayonove, C. L. (1998). Effect of sun exposure on carotenoids and C13-norisoprenoids glycolises in Syrah berries (*Vitis Vinifera* L.). *Sciences des Aliments*, *18*, 361-373.

[64] Symington, C., Ferreira, A. & Rogerson, F. (2011). Industrial trials modulating Touriga Nacional aroma tipicity. *Proceeding from the World Congress of Vine & Wine*, Porto, Portugal, *34*.

[65] Carvalho, E. S., Correa, L. C., Camargo, A. C. M. & Lima, A. (2018). Phenolic profile of 'touriga nacional' wine: influence of maceration time during the winemaking. *Proceedings from the Simpósio Latino-Americano de Ciência dos Alimentos*, Campinas, Brazil. 12. Recovered from: https://proceedings.science/slaca/slaca-2017/papers/phenolic-profile-of----touriga-nacional----wine--influence-of-maceration-time-during-the-winemaking-?lang=pt-br.

[66] Langbecker, M. R., Eckhardt. D. P., Cunha, W. M., Costa, V. B., Gabbardo, M., Schumacher, R. L. & Andrade, S. B. (2018). Experience of Cold Maceration on "Touriga Nacional" Wine Varieties in the Campanha Gaúcha Region, Brazil. *Journal of Experimental Agriculture International*, *22*(3), 1-8.

In: Fermented and Distilled
Editors: M. B. M. de Castilhos et al.
ISBN: 978-1-53618-985-8
© 2021 Nova Science Publishers, Inc.

Chapter 6

TANNAT WINE: CHARACTERISTICS AND KEY STAGES IN ITS PRODUCTION

Laura Fariña[1], Karina Medina[1], Valentina Martín[1], Francisco Carrau[1], Eduardo Dellacassa[2] and Eduardo Boido[1,*]

[1]Área de Enología y Biotecnología de Fermentaciones, Departamento de Ciencia y Tecnología de los Alimentos, Facultad de Química-UdelaR, Montevideo, Uruguay

[2]Laboratorio de Biotecnología de Aromas, Departamento de Química Orgánica, Facultad de Química-UdelaR, Montevideo, Uruguay

ABSTRACT

Since the '90s Uruguay has aimed at producing quality Tannat wines, as a way of making its wine industry known in the world. Supporting this development from our research group as well as from other institutions we have collaborated to the understanding of the key processes in the production of quality Tannat wines as well as their distinctive characteristics. In this chapter are summarized the main advances in the

---

[*] Corresponding Author's E-mail: eboido@fq.edu.uy.

study of each biotechnological stages that lead to the elaboration of a Tannat wine as well as their effect on the primary metabolites responsible for its sensory quality (polyphenols and aromas). We also present the perspectives of our work on which we will focus our research in the coming years.

Keywords: polyphenols, anthocyanins, flavanol, aroma, aroma precursors, sensory properties, aging

INTRODUCTION

Tannat is the red grape *Vitis vinifera* introduced in Uruguay in the nineteenth century by French Basques emigrants. It was so well adapted to our environmental conditions that it is considered the grape for the production of the flagship wines of the country. Twenty years ago, Uruguay was considered one of the few places in the world where Tannat was grown, although its origin was from the France Southwest. In regions such as Bearn, Madiran or Irouleguy, it was not so popular as a varietal wine. It is usually blended with Cabernets due to its rich tannin and color contribution, its primary characteristics. It indicates from where its name derived: tannin and tanned or highly colored. Today this variety was planted in many other wine regions such as California, Argentina, or New Zealand [1].

In 1994, our group of enology, both chemists and biologists, traced a research program and started to employ the most efficient analytical methods to conduct their research on our primary grape, the Tannat [2]. New molecules of grape and wine color, aroma, and flavor have been identified [3], and in 2012 Tannat was the first commercial cultivated grapevine whose genome had been sequenced with high coverage and with significant contributions to the understanding of the polyphenol metabolic pathways in plants [4]. Sensory analysis, included in the laboratory alongside chemical analysis methods, reveals the importance of molecules present at very low concentrations and the importance of the interactions between them. Wondering about how chemical composition affects the flavor and body of wine took our group from the work in viticulture and enology into grape and yeast metabolomics research during vinification [5–6]. This knowledge always should be developed considering the "low input winemaking" philosophy to reduce the grape and wine manipulations for increasing flavor complexity and differentiation [7].

The strategy is to predict the quality of the grape as a result of genetic indicators with an impact on aroma and Tannat polyphenols [8]. The correlation between the "terroir" conditions with gene expression of key quality metabolites of the Tannat grape berry development and the final wine, such as procyanidins and anthocyanins within polyphenols and norisoprenoids regarding flavor precursors was significantly relevant for Tannat wine. Identifying genes whose gene expression significantly increases these compounds during the various stages of berry maturity development is one of our research targets today. This challenge will allow us to improve many technological aspects of Tannat wines such as differential vineyard management, gene expression in different Tannat clones, soil region effects, native yeasts, and plant interactions, wine fermentation, and aging. In this chapter, we make a summary of all these aspects that have been advanced about Tannat grape and wine, which is now one of the best known red grape varieties for winemaking.

POLYPHENOLS IN GRAPE AND WINES

The quality of plant-derived food products and wine is closely related to phenolic compounds. They are responsible for the color of the red grapes and wines, and they are involved in the oxidative browning of white wines. They also contribute to taste and astringency as a result of the interactions with salivary proteins [9]. Phenolic compounds in grapes can be found primarily in the skin and seeds of grape berries, and they can be classified as nonflavonoids (hydroxybenzoic and hydroxycinnamic acids and derivatives, and stilbenes) and flavonoids (anthocyanins, flavan-3-ols, flavonols, and dihydroflavonols).

Grapes contain non-flavonoid compounds mainly in the pulp, while flavonoid compounds are located in the skins, seeds, and stems [10]. Their composition is affected by several factors, such as grape variety, ripening stage, climate, soil, vineyard location, and cultivation. These compounds have also shown to be crucial chemical markers to characterize different varieties of grapes, especially anthocyanins [10–11]. The phenolic potential, mainly anthocyanins, total pigment content, and their variations with vintage and vineyard treatments, have also been reported for Tannat grapes [12–15], as well as the anthocyanin and flavanol contents, and antioxidant capacity of Tannat wines at different ages [12, 13, 16-18].

Generally, grape and wine phenolic analysis can be divided into three fundamental areas: development of phenolic compounds in the vineyard, extraction, and modification of phenolic compounds during wine production and transformation during aging (Figure 1).

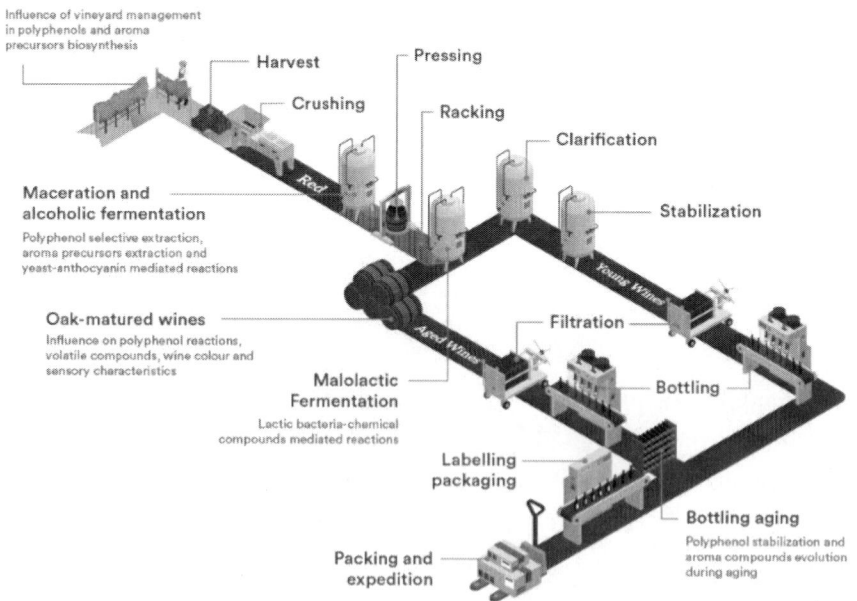

Figure 1. Scheme of vinification for the production of young or aged Tannat wine.

Grape and wine phenolic research may also identify and determine the level of the different phenolic compounds present in wine, evaluating their sensory importance and interaction with other compounds to make the information useful for the wine industry.

Anthocyanins in Tannat Grapes and Wines

Anthocyanins are the pigments responsible for the grape color, which are extracted to the must during winemaking, providing the characteristic purple-red hue of young red wines. Once extracted, and until the end of the fermentation, its concentration begins to decline probably due to their adsorption by yeasts in the first steps of winemaking. Later, the concentration decreases due to the reactions of condensation, polymerization, oxidation, and precipitation, which could be involved in anthocyanin disappearance. Some of these reactions imply the degradation of the anthocyanin, whereas others yield products that can provide, in turn, different hues to the wine.

One of the most remarkable features of Tannat wines is their intense color resulting from its highest pigment contents when compared to red wines produced from other grape varieties [13, 16, 19]. The international wine market shows an increasing demand for Tannat wines due to their originality and excellent quality [3].

Tannat grapes were characterized as presenting high concentrations of anthocyanin in skins, with an increase in its concentration during ripening. The highest levels of anthocyanins found were malvidin, delphinidin, and petunidin monoglucosides. The ratios of peonidin/cyanidin and malvidin/delphinidin in Tannat grapes increased significantly during ripening, indicating differences in the biosynthetic rate for these compounds that may be caused by the differences in activity or expression of *O*-methyl transferases (OMTs). In ripe berries, there was a predominance of blue tri-hydroxylated anthocyanins (90%), with no change during maturation [15].

Extraction and Loss of Anthocyanin during Vinification

In traditional winemaking, only 40% of the grapes' anthocyanins are transferred to the wine [20]. The limited extraction of anthocyanins is primarily due to the lack of permeability of cell walls and cytoplasmic membranes [10–21] because these compounds are in the skin, in the upper cellular layers of the hypodermis. The composition of cell walls is genetically determined and modifies the changes in the hardness of skin and seed tissues along with ripening. The simultaneous development of maceration and alcoholic fermentation influences the extraction of polyphenols due to the ethanol content that provokes the disintegration of the vacuolar membranes and the walls of the skin cell [22]. Anthocyanins are compounds easily soluble in water and, therefore, they are in solution from the beginning of the maceration, independent of the ethanol concentration [23].

Several alternative techniques for maceration have been proposed to produce a focused transference of the phenolic and aromatic compounds of the grape to the wine, and so improving quality and aging potential [24–25]. Most of these techniques have had a substantial impact on the color of red wines [25]. Hot maceration improved the intensity and quality of the Tannat wine color by increasing the extraction of phenolic compounds and, at the same time, promoting condensation between anthocyanins and tannins, giving better color stability [26].

At the fermentation step, yeasts have been demonstrated to interact with anthocyanins in different ways. The yeast cell-wall material producing anthocyanin adsorption is a well-known phenomenon [27–29]. The different strains of *Saccharomyces cerevisiae* showed higher percentages of anthocyanin removal with increasing anthocyanin polarity [29]. For Tannat, the average losses of individual anthocyanins in red juice medium supplemented with anthocyanin extract of Tannat grape skins [29], were delphinidin, 45 to 50%; petunidin, 22 to 30%; peonidin, 7 to 27%; malvidin, 10 to 15%; and acylated anthocyanins, 9 to 16%. The work of Medina et al. [29] showed no correlation between color intensity and total anthocyanin concentration after fermentation, due to the formation of anthocyanin derivatives. The color intensity of these pigments at the wine acidic pH is higher than of the anthocyanins, which exist essentially as colorless hemiacetal structures in wine [30].

The interaction of non-*Saccharomyces* with anthocyanins during vinification was studied by Medina et al. [31], who demonstrated that color intensity, hue, and total anthocyanin were significantly affected by different strains of *Metschnikowia pulcherrima*, *Metschnikowia fructicola*, *Hanseniaspora vineae*, *Hanseniaspora uvarum*, *Hanseniaspora opuntiae*, *Hanseniaspora clermontiae*, *Hanseniaspora guilliermondii*, *Torulaspora delbrueckii*, *Starmerella bacillaris*, *Issatchenkia terricola*, *Candida railenensis*, and *Candida shetae*. The formation of anthocyanin derivatives vitisin A, vitisin B, and vinylphenols adducts by the yeast genera *Hanseniaspora* and *Metschnikowia* was also reported [31].

Moreover, the non-*Saccharomyces* and *Saccharomyces* strains co-fermentation in Tannat wines increase acetaldehyde accumulation with a simultaneous increase in anthocyanin derived pigment formation [32].

Modification of Anthocyanin during Aging

Boido et al. [18] studied the pigment composition of different Tannat wines produced in Uruguay in consecutive vintages to determine the qualitative and quantitative changes produced in the principal pigment families (anthocyanins, pyranoanthocyanins, direct and acetaldehyde-meditated flavanol-anthocyanin condensation products) as wine became older. The determination of the color properties of the Tannat wines (by calculation of the CIELAB parameters) was also considered to monitor the color evolution and to establish the contribution to the wine color of the different pigment families at different stages of wine life [18].

The anthocyanin family, which accounted for 98% in grapes, represented only 32% of the total quantified compounds in the 64 months old Tannat wines, although the higher decrease was observed from the 40-month sample (Figure 2). In contrast, the percentage of pyranoanthocyanins, over the total, increased with age, representing 54% in the oldest sample. In this family, the vinyl adducts (4-vinylphenol and 4-vinylcatechol adducts) showed a substantial increase (33% of the total area in the 64-month-old sample). The percentage of the pyranoanthocyanins increased not only by their higher stability compared with that of the anthocyanins, but also due to some of the members of this family were synthesized during all of the study period. Thus maintaining their concentration in the time concerning that of the anthocyanins and to the total pigment content [33].

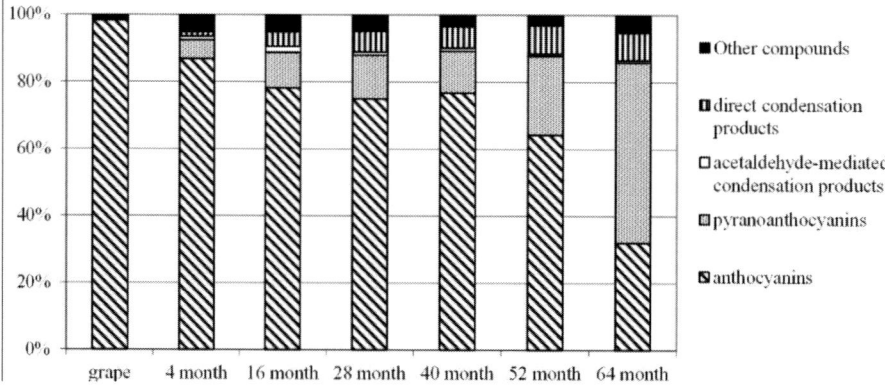

Figure 2. Evolution of the percentages of the different pigment families from the grape to 60 months aged Tannat wines.

The direct condensation products between flavanols and anthocyanins showed an increasing relevance as Tannat wine aged, representing more than 8% of the total area in the oldest sample versus 0.1% in grapes and 1.4% in the youngest wine sample. In contrast, acetaldehyde-mediated condensation products decreased significantly after the maximum during barrel aging, remaining almost constant at low levels (almost 0.8% of the total pigment content). Thus, the work of Boido et al. [18] showed that, during wine aging, the pyranoanthocyanins and, to a lesser extent, the flavanol-anthocyanin direct condensation products acquire increasing importance in quantitative terms, reporting the pyranoanthocyanins as the most abundant pigments in the oldest analyzed sample (64 months old). Furthermore, at wine pH, most of these anthocyanin derivatives are present in colored forms, whereas only 15% of the anthocyanins are in flavylium colored form [35], which would increase the importance of the anthocyanin derivatives in the color of the wine during aging.

In brief, the decrease in the Tannat wine anthocyanin levels would allow the expression of the anthocyanin derivatives color: pyranoanthocyanins and direct condensation products during the three first years of aging, and essentially pyranoanthocyanins in the last 3 years.

Flavanol and Other Polyphenolic Compounds in Grapes and Tannat Wines

The different polyphenolic families (phenolic acids, flavonols, and flavanols monomers and oligomers) present in Tannat grapes skins and seeds was studied in detail by Boido et al. (2011) [15] to establish qualitative and quantitative changes in these compounds during grape ripening and vinification. The results also enabled to establish a characteristic polyphenol profile for Tannat young wines.

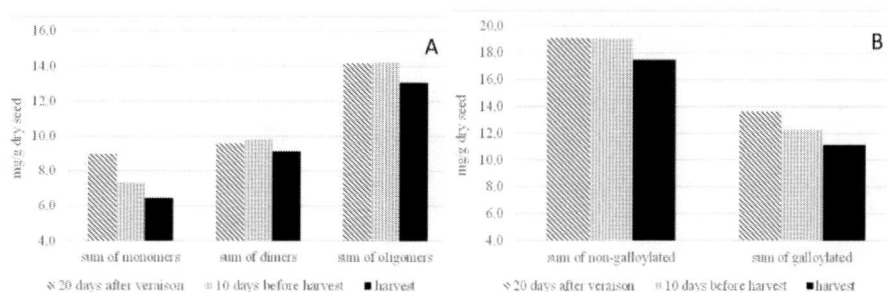

Figure 3. Evolution of flavan-3-ol content in seeds during ripening of Tannat grapes.

This study showed that Tannat seeds had a higher content of flavan-3-ols than that reported for many other grape varieties. The content of these compounds (expressed as mg/g of dry seed) declined during maturation in agreement with data obtained for other red grape varieties [35–36]. The percentage of monomers decreased (Figure 3), indicated an increase in the mean degree of polymerization [15]. Galloylated compounds represented almost 40% of the total flavan-3-ols, which decreased during ripening as a result of a corresponding decrease of epicatechin gallate.

Total flavan-3-ols in skins showed no significant differences during ripening, with values similar to those found in other red grape varieties. The flavan-3-ols profile in skins was characterized by a very low content of galloylated forms, the presence of monomeric forms as the most abundant with a decrease in their percentage during ripening, and a 30–35% prodelphinidin content.

Epicatechin gallate was not detected in Tannat wine, and galloylated forms represented a low percentage of the total flavan-3-ols, indicating ineffective extraction or hydrolysis of these compounds during vinification [15].

In Tannat grapes, the principal flavonols were quercetin and myricetin aglycones in their galactoside and glucoside form, as in a large number of other red grape varieties [37].

Eleven phenolic acids were identified in Tannat grapes and wines, but only gallic and protocatechuic acid were found in seeds. These acids increased their concentration in seeds during ripening; however, a corresponding increase in these phenolic acids in the skins was not evident. The hydroxycinnamic acid content was higher than the hydroxybenzoic acid in Tannat skins, representing 75% of the total phenolic acids at harvest. The concentration of trans-caftaric and trans-coutaric acids reported in Cabernet Sauvignon, Tempranillo, and native varieties of Andalucia, was lower than in the Tannat variety [38].

In Tannat wines, the phenolic acid content was higher than the grapes [15], possibly caused by a contribution of phenolic acids present in the pulp [39].

Sensory Properties

Astringency is one of the most essential sensory characteristics that define the complexity and quality of red wine [40]. It is a tactile sensation caused by the interaction of polyphenolic compounds and salivary proteins, which leads to a decrease in the lubrication of the oral epithelium [41]. Astringency involves several mouth-feel sensations, which have been commonly used to describe red wines, and unlike taste sensations, astringency perception is strongly time-dependent [42].

Vidal et al. [43] described the astringency of forty commercial Uruguayan Tannat wines using a wide range of astringency sub-qualities, from soft related textures such as silky, velvety, and suede, to those related to excessive astringency, such as harsh, hard, and aggressive. Four groups of samples with different astringency characteristics were identified, but this sorting was not related to vintage, price segment, or aging in oak barrels [43]. Using Temporal Dominance of Sensations (TDS) [44] characterized the dynamics of wine astringency identifying significant differences in Tannat wines, which were not identified using static methods. Astringency intensity was not significantly correlated to the dominance of astringency terms, while wines that did not significantly differ in their average astringency intensity showed different dynamic astringency profiles.

The relationship between astringency and phenolic composition of commercial Uruguayan Tannat wines was performed using the novel predictive method boosted regression trees [45]. Boosted regression trees technique takes advantage of statistics and machine learning techniques, combining a large number of simple regression trees building a single model that optimizes the predictive performance. Besides, they can model nonlinear responses, which are likely to be relevant for astringency perception.

In the study of Vidal et al. (2018) [45], flavan-3-ols were the family of phenolic compounds with the highest contribution to astringency intensity and astringency sub-qualities. Also, some dimers, trimmers, and the sum of non-galloylated tetramers were consistently selected as relevant contributors to some astringency sub-qualities.

Polyphenols and Health

One of the factors that have promoted research interest in wine polyphenols was the epidemiological study of Serge Renaud on the so-called "French paradox", which found a lower incidence of coronary heart diseases in France as a consequence of the regular drinking of wine [46]. Accordingly, polyphenols acquired from the moderate consumption of red wine also provide pharmaceutical and nutritional benefits to humans [47–48], such as helping to prevent cancer and reducing the inflammation associated with coronary artery disease [49]. Proanthocyanidins are the principal vasoactive polyphenols in red wine, they induce the endothelium-dependent dilatation of blood vessels and suppress the synthesis of the vasoconstrictive peptide endothelin-1 [50–52].

Briefly, the particular profile of phenols in Tannat wines contributes to health benefits as antioxidants preventing cardiovascular diseases and some types of cancer, adding value to this particular variety, which represents the emblematic *Vitis vinifera* variety for Uruguay.

WINE AROMA

Food consumption involves a multimodal experience that sensory information is processed through several inputs (olfactory, tactile, and visual), and information is combined, providing an overall sensation that is best defined by the term "flavor" [53].

Aroma compounds that comprise food flavors occur in nature as complex mixtures of volatile compounds. Flavor chemists have worked for a long time to elucidate the identity of pure aroma chemicals that possess the unique flavor characteristics of the food from which they were derived. Usually, these unique flavor compounds are referred to as "character-impact compounds" [54]. Often, character impact is elicited by a synergistic blend of several aroma chemicals. The total concentration of these naturally occurring components varies from a few parts-per-million (ppm) to approximately 100 ppm, with the concentration of individual compounds ranging from parts per billion to parts per trillion [55]. However, many of the volatile chemicals that have been isolated from natural flavor extracts do not elicit aroma contributions that are reminiscent of the flavor compound by itself. Compounds that are considered as aroma substances are primarily those that are present in concentrations higher than their respective odor thresholds. Compounds with concentrations lower than the odor thresholds also contribute to aroma when mixtures of them exceed these thresholds.

Varietal aromas come from the grapes, constitute their aromatic potential, and they are responsible, principally, for the typical characteristics of the wine. They can be divided into free aromatic compounds (terpenes and pyrazines) and aroma precursor compounds (glycosidic compounds, carotenoids, cysteine compounds, and phenolic acids). Aroma precursors are compounds that are not odoriferous in grapes, but under certain circumstances, they may become volatile and participate in the aroma. In the winemaking process, these molecules are transferred to the wine both as free and bound forms [56], its profiling and content in grape products depends on the grape variety and how it interacts with berry ripeness and climatic and agronomic factors [57].

Management of Aroma Precursors in Tannat Grapes

In Tannat grapes, the presence of components in the free state is negligible, and the significant contribution to the aroma of wine comes from its aroma precursors. In this sense, in Tannat variety, phenols and norisoprenoids represent almost 80% of the total glycosidic components. Terpenes have been reported as possessing an essential aromatic role in other *Vitis vinifera*, but in Tannat, they represent only 13% of the aroma fraction. Figure 4 shows the importance of the different groups of varietal glycoside compounds [58].

Viticulture and vineyard management practices have effects on the quality of the grapes and the wine. In Tannat grapes, the impacts of some cultural practices on fruit composition were evaluated with emphasis on the biosynthesis of aroma precursors.

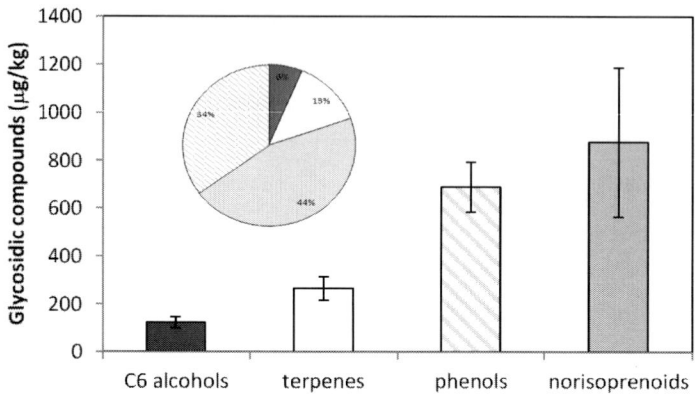

Figure 4. Importance of different groups of Tannat varietal glycoside compounds.

Pruning is a key stage, and different systems can be used, with different results on the quality of the grapes and the wines obtained. In the Tannat variety, the effect of two pruning systems widely used among Uruguayan winegrowers was studied by Fariña (2008) [59] and their effect on glycosidic compounds, and classic grape quality parameters were evaluated. Using trellis conduction, cane and spur pruning systems were evaluated during ripening in three consecutive harvests. Table 1 shows the average result obtained on the date of harvest of the three years analyzed. The load (kg/plant) obtained at the time of harvest was considered when estimating the differences between the pruning systems. Grapes from cane pruning presented a significantly higher concentration of glycosidic terpene precursors and norisoprenoids [59]. A limitation when evaluating the optimal harvest date is the complexity of the analysis of the aromatic precursors. In this experience, the correlation with other classic grape quality parameters (sugar, pH, acidity, grain weight, total anthocyanins, total polyphenols) was evaluated, but this correlation was not established (Figure 5) [60]. Therefore, the aromatic maturity of the Tannat grapes can only be assessed by monitoring this parameter. The development of easy and accessible methods such as the use of Near-Infrared (NIR) technology is feasible to efficiently assess the grape ripening process for vintage decisions [58].

Table 1. Terpene and norisoprenoid content determined in Tannat grapes at the date of harvest

	3rd harvest		2nd harvest		1st harvest		ANOVA		
	Cane	Spure	Cane	Spure	Cane	Spure	Harvest (H)	Pruning (P)	Interaction: H/P
	Average (mg/L) ± SD								
terpenes	1016 ± 171	571 ± 63	293 ± 72	114 ± 14	378 ± 67	182 ± 48	**	*	ns
norisoprenoids	751 ± 208	384 ± 180	161 ± 80	32 ± 4	875 ± 440	235 ± 29	**	**	ns

* p≤ 0,1; **p≤0,05; ns: no significant

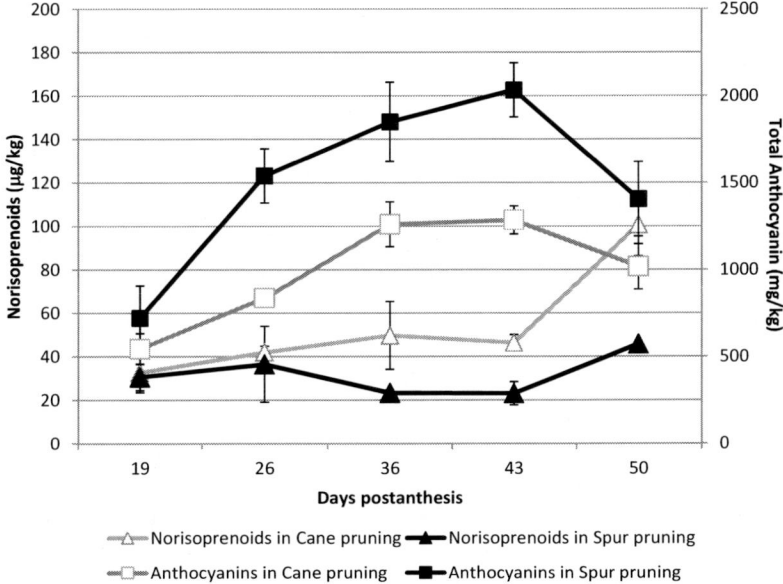

Figure 5. Norisoprenoid and total anthocyanins content evolution during Tannat grapes ripening.

As mentioned above, C13-norisoprenoids are relevant aromatic compounds in Tannat grapes. These compounds can be formed from glycosylated intermediates and by direct degradation of carotenoids such as β-carotene, lutein, neoxanthin, and violaxanthin. Carotenoids are photosynthetic pigments, which are very susceptible to degradation when exposed to light, oxygen, moisture conditions, and high temperatures. Aiming at evaluating the importance of the light exposition of Tannat vineyards on aroma precursors, the effect of luminosity on the profile of carotenoid, and their degradation products in the Tannat grapes was studied for two consecutive years [33]. The difference in luminosity was established by applying a stone mulch soil (SMS) in a vineyard, and its effect was compared with the bare soil (SB). Higher carotenoid concentrations were found in the vineyard with SMS than in the vineyard with SB throughout both ripening seasons. In the initial sampling, the total carotenoid concentration was higher in the vineyard with SMS than in the vineyard with SB (108% and 55% respectively) in the two years analyzed. Total carotenoid concentration decreased steadily towards maturity: 46% in the SMS and 41% in the SB during the first year and 23% and 8% in the second year (SMS and SB, respectively) [33].

Figure 6 showed the evolution during the ripening of C13-norisoprenoids and carotenoids during the second year of analysis. The results were consistent with fruit light exposure data (measured in photosynthetically active radiation), but they did not correlate with temperature measured in the fruit [61].

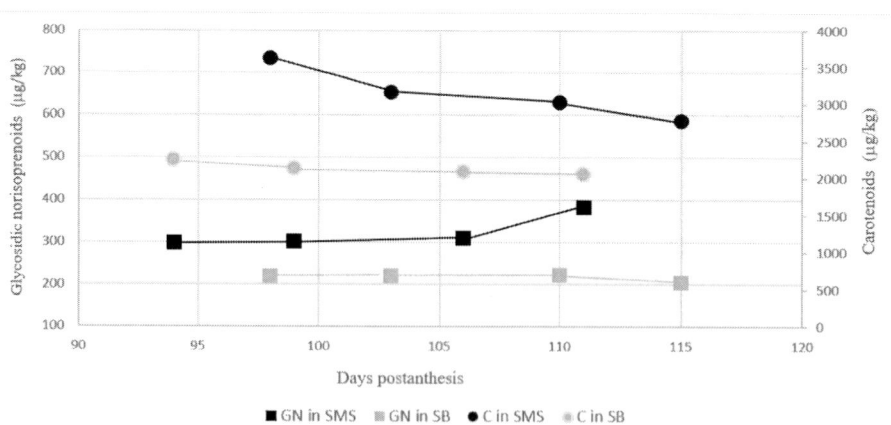

Figure 6. Evolution of norisoprenoids and carotenoids in Tannat grapes in two different sunlight and luminosity conditions (GN = glycosidic norisoprenoids and C= carotenoids).

Additionally, as genetic grape variety origin influences its metabolic expression, different clones of the same grape variety can have different production characteristics and give rise to wines with different sensory properties [62]. In the 1990s, certified materials (with clonal selection) of the Tannat variety selected in France were introduced in Uruguay. From that moment on, clone 398 was consolidated as the most cultivated, but without having any information about its qualitative attributes in our territory [63].

In the last few years, Fariña et al. (2018) [64] have conducted a quali-quantitative study of the aromatic grape precursors present in the commercial Tannat clones vineyards (clones 398, 399, 472, 474, 475, 717, 794, 944) to identify those clones that stand out from the aromatic point of view. The results indicated that clones 717 and 475 presented a significantly higher level of glycosylated norisoprenoid precursors followed by clones 399 and 398. The importance of this research is based on the role that these precursors play in high-end wines, as further explained in this chapter.

Impact of Varietal and Fermentative Aromas on Young Tannat Wines

The flavor of the wine originates from the grapes. The treatment of the must (grape juice), the fermentation, and the maturation process of the wine are directly associated with the chemistry of the entire winemaking process, resulting in the unique wine taste and smell.

As mentioned above, the secondary metabolism expression in grapes is responsible for the principal aroma compounds in grape must, providing the basis of the wine varietal character [33]. Fermentation increases the chemical and aroma complexity of wine by assisting in the extraction of compounds from solids present in the grape must, modifying some grape-derived compounds, and producing a substantial amount of yeast metabolites [65].

During the vinification process, Tannat grapes release a wide variety of volatile organic compounds from the entire plant organs, and their profile is primarily comprised of esters, alcohols, aldehydes, ketones, lactones, terpenoids, and carotenoid derivatives [33]. The release of bound volatile compounds can be achieved by physical (temperature), chemical agents (acidification), and biochemical methods (enzymes), which are the most effective [67]. Upon glycoside hydrolysis - by the action of heat, acid hydrolysis, or enzymes - the corresponding aglycones (free forms) will be volatilized [68]. At this step, the role of yeast is crucial, due to both the *Saccharomyces* [68] and non-*Saccharomyces* yeasts [69-72] ß-glycosidase production.

Among the compounds found in Tannat young wines, the ratio between *trans*- and *cis*-3-hexen-1-ol contents and the relationships between them can be considered as characteristic of this *V. vinifera* variety [67, 73], considering *cis*-3-hexen-1-ol content higher than *trans* form.

Monoterpene compounds (free stage) were all found under their own thresholds, as usual for wines from neutral cultivars [74]; however, in some cases, rather high contents of limonene and geraniol were found.

When glycoconjugates (bound forms) were studied in young Tannat wines, C13-norisoprenoids made up 38% and monoterpene volatiles 7% of the total level of the volatiles released, while the other volatiles were C6 alcohols (4%) and benzenoid compounds (52%) [66].

In a typical Tannat wine fermentation, *Saccharomyces cerevisiae* is used to carry out the process. The yeasts used contribute to the aroma, not only regarding the contribution of the produced enzymes but primarily by their secondary metabolism. Concerning the fermentation compounds, Tannat wines contain remarkable amounts of 2-phenylethanol and low contents of higher alcohol acetates, as well as fatty acid ethyl esters [75]. Although the concentrations of alcohols and their esters in wine are determined essentially by the yeast strain and the conditions of fermentation [76], it is relevant to note that amino acids are involved in the biosynthesis pathway of these compounds [77]. Thus, the profile and concentration of alcohols and esters in wine are also affected by the grape variety.

When Tannat wine is elaborated from non-*Saccharomyces* yeasts or by mixed culture between *Saccharomyces* and non-*Saccharomyces*, higher levels of acetates and ethyl esters (fruity notes) are found with higher levels of linear alcohols, which contribute to increasing the aromatic quality of Tannat young wines [78].

In particular and aiming at studying the effect of non-*Saccharomyces* on the volatile compounds produced during Tannat red grape vinification, Medina et al., [78] conducted an experiment using pure cultures of *Saccharomyces cerevisiae*, and mixed cultures by sequential inoculation of *Hanseniaspora vineae* (T02/05F), and *H. clermontiae* (A10/82F), with a *Saccharomyces cerevisiae* conventional strain. The vinification systems applied to Tannat grapes were conducted at three production scales: semi-pilot (20 kg), pilot (500 kg), and industrial scale (5000 kg). Fifty-one volatile compounds were identified in all the vinifications, sixteen of them above the aroma threshold values, contributing potentially to the final sensory profiles. The highest formation of acetates was detected in the vinifications conducted with *H. vineae*, whereas the maximum ethyl acetate concentration occurred in the vinification with *H. clermontiae*. Interestingly, the highest concentration of norisoprenoid compounds was achieved by *H. vineae* vinification at industrial-scale compared to micro-fermentations [79]. In conclusion, it was demonstrated that Tannat wines resulted in wines with improved sensorial profile, where industrial vinifications produced wines described as "fruity" and "woody" [78].

Malolactic fermentation (MLF) is another required stage in vinification, primarily for red wines elaboration. Malolactic bacteria may influence wine aroma and flavor by various mechanisms, including the production of volatile secondary metabolites and the modification of grape and yeast-derived metabolites. The effect of MLF on volatile compounds in Tannat wines was studied by Boido et al. (2009) [80]. The results obtained can be summarized as follows: the concentration of ethyl lactate increased, but this concentration depended on the strain of *Oenococcus oeni* used. The concentration of other esters such as isobutyl, isoamyl, hexyl, and 2-phenylethyl acetate, and ethyl hexanoate (which contribute to pleasant fruity notes) changed after MLF and changes are strain-dependent.

Some strains of *O. oeni* increments the contents of diethyl succinate and γ-butyrolactone during MLF of Tannat wine. Alcohols had similar behavior, i.e., several alcohols presented higher concentration after MLF (2-phenylethanol and 3-methylthio-1-propanol), but the sum of alcohol content showed no significant differences, also considered as *O. oeni* strain dependent [80].

Evolution of Tannat Wine Aroma during Aging

The aging process of wine comprises a combination of stages beginning with the maturation in oak barrels and continuing in the bottle (Figure 1). Aging has a remarkable effect on the wine volatile compounds and, finally, on its aroma, which is significantly improved. The aging impact on wine aroma has been used to introduce variations in winemaking. On some occasions, MLF can take place in the oak barrel; on other situations, wines only undergo a process of aging in the bottle (without previous maturation in oak barrels).

During maturation in oak barrels, Tannat wine acquires typical notes by extraction of components from the wood that plays a crucial role in the wine "bouquet".

When Tannat wine was submitted to alcoholic fermentation in tanks, and then transferred to oak barrels where MLF was developed, the aroma data yielded by the wine as a result of the extraction of volatile compounds from the barrel during a maturation period of 540 days is shown in Table 2 [59]. Wine MLF fermentation and aging in bottles were also performed in parallel. For comparison purposes, the results of control wine are also presented in Table 2, showing the impact of aroma components from the oak extracted during the MLF [59].

Table 2. Volatile compounds extracted from oak barrels during a maturation time of 540 days of Tannat wine

	oak barrel				bottle*		
	Time (days)						
	0	180	360	540	540	Odor Descriptor	Odor threshold (µg/L)
	µg/L ± D.M.	µg/L ± D.M.	µg/L ± D.M.	µg/L ± D.M.	µg/L ± D.M.		
furfural	n.d.	72 ± 50	35 ± 9	31 ± 6	n.d.	almond	20000
furfuryl alcohol	1159 ± 224	1025 ± 218	1788 ± 370	1097 ± 254	619 ± 39	musty	45000
5-methyl-furfural	n.d.	49 ± 17	84 ± 18	95 ± 13	n.d.	rosted almond	45000
5-hydroxy-methyl-furfural	n.d.	186 ± 21	282 ± 6	161 ± 61	n.d.	almond	100000
furfuryl ethyl ether	n.d.	10 ± 6	52 ± 8	38	n.d.	solvent, kerosene	430
5-methyl furfuryl ethyl ether	n.d.	n.d.	31 ± 13	21	n.d.	not reported	1000
cis-whisky lactone	38 ± 25	77 ± 9	127 ± 13	155 ± 12	72 ± 22	coconut, oak	46
trans-whisky lactone	54 ± 39	110 ± 6	195 ± 50	239 ± 31	140 ± 39	coconut, oak	32
maltol	n.d.	17	58 ± 29	53 ± 3	7 ± 3	sweet, caramel-selling	5000
guaiacol	178 ± 7	17 ± 2	23 ± 4	188 ± 55	230 ± 69	sweet smoky	75
4-propyl-guaiacol	n.d.	18	19 ± 7	20 ± 8	n.d.	leather, animal	10
syringaldehyde	n.d.	n.d.	n.d.	265 ± 41	n.d.	spaces, smoke	50000
vanillin	n.d.	112 ± 16	170 ± 1	147 ± 13	n.d.	vanilla	320
vanillic acid	n.d.	139 ± 12	260 ± 31	126 ± 39	n.d.	vanilla	25000
ethyl vanillate	n.d.	141 ± 10	215 ± 59	82 ± 22	n.d.	vanilla	990
acetovanillone	n.d.	n.d.	271 ± 20	270 ± 50	190 ± 3	wood, vanilla	25000
2,6-dimethoxyphenol	n.d.	n.d.	99 ± 19	612 ± 159	527 ± 86	smoke, medicinal	57
4-allyl-2,6-dimethoxyphenol	n.d.	18 ± 4	36 ± 4	54 ± 4	24 ± 5	phenolic, smoke	1200

*This wine was transferred to the bottle after the MLF in barrel.

The main change in the volatile fraction of Tannat wine during aging is the decrease in the level of aliphatic ethyl esters and acetates [59, 80]. Boido et al., [80] reported an increase in the levels of γ-butyrolactone, diethyl succinate, ethyl succinate and diethyl 2-hydroxyglutarate during aging (also reported for other varieties). During the aging stage, ester formation is also achieved through chemical esterification.

Aging also shows an effect on varietal components of Tannat wines, as small amounts of free-state norisoprenoids β-damascenone, α-ionone, β-ionone, 3-oxo-α-ionol; riesling acetal and vomifoliol were also identified. The concentrations reached by some of these norisoprenoids in aged Tannat wines exceed the individual perception threshold [73–75], enhancing fruity, dry raisin, or red plum notes (depending on the concentration) [75].

Aroma Sensory Characterization

Descriptive analysis with highly trained panels has been the most widely used methodology for characterizing the aromatic profile of wines [81–83]. In this methodology, assessors are trained in the identification and quantification of specific notes to provide a qualitative and quantitative description of wine aroma [84]. The analysis of Tannat young wines allowed associating them with the secondary and tertiary descriptors berry (blackcurrant), dried fruit (prune), resinous (oak), spicy (liquorice), and microbiological (yeasty) [85]. MLF exerts a significant decrease of secondary descriptors 'berry fruit' and 'fresh vegetative' in Tannat wines, as well as a decrease in related tertiary descriptors such as 'blackcurrant', 'apricot', 'cut green grass', and 'green pepper'.

In contrast to the descriptive analysis, holistic methodologies rely on the evaluation of global similarities and differences among samples, encouraging the generation of a synthetic representation of the products, which is inhibited when assessors are asked to focus their attention on specific characteristics [86, 87]. Projective mapping is a holistic methodology for sensory characterization, in which consumers are asked to provide a two dimensional projection of a group of samples, according to their criteria [88]. In the aroma profile of Tannat wine samples, the primary descriptors used were red fruit, fruity, dry fruit, and woody. These results are in agreement with the previously reported by the descriptive analysis technique [85].

CONCLUSION AND PERSPECTIVES

Tannat wine represents the emblematic *Vitis vinifera* variety of Uruguay, in which international demand shows an increase because of its originality and quality. This variety is well known for its aroma profile reminiscent of red fruit, fruity, dry fruit, and woody, high contents of tannins considered one of the highest levels of phenolic compounds reported for *Vinifera* grape varieties and intense color concentration. The high grape maturity and intense barrel aging over the lees contribute to the astringency reduction and improve body and sensory balance to the final wine.

Another interesting fact related to the high phenolic content is the contribution of Tannat wines as antioxidants in a healthy diet since several studies are focus on the understanding of its contribution to preventing cardiovascular diseases, cancer, and in general, extending cell life span.

In terms of varietal aromas, the main contribution is given by the aromatic precursors since free aromas are in very low concentration. The current study of the metabolic pathways involved in the biosynthesis of these aromatic precursors will allow the development of strategies at the level of viticulture and vineyard management practices to promote the production of these essential metabolites in the future.

Yeast and lactic bacteria performance during fermentation have crucial effects on increase concentration of free aroma compounds and the biosynthesis of some key aroma precursors, improving the whole flavor characteristics of the final wine.

Concerning the reported incidence of non-*Saccharomyces* yeasts on the aromatic profile of wines and the relevant diversity of native yeasts found in Tannat grapes, studies have been carried out to use alternative yeasts to enhance the aromatic profile of this grape variety. It was demonstrated the potential of the *Hanseniaspora vineae* as one of the most appropriate non-*Saccharomyces* yeasts to provide this effect in Tannat wine due to its high capacity to produce derived amino acids acetates, in particular 2-phenylethyl acetate and benzyl alcohol. Commercial wines are been produced with this yeast at a winery scale with preliminary suitable sensory evaluations with consumers and winemakers that confirm sensory results obtained on a smaller scale.

In summary, the information presented in this chapter helped to understand the phenolic and aromatic winemaking potential of Tannat variety and its functional properties within food chemistry.

REFERENCES

[1] Harding, J. Tannat: home away from home - World Of Fine Wine. (2018). *World Fine Wine*. Recovered from: http://www.worldoffinewine.com/news/tannat-home-away-from-home-7012002.

[2] Carrau, F. M. (1997). The emergence of a new uruguayan wine industry. *International Journal of Phytoremediation, 21*, 179-185.

[3] Carrau, F., Boido, E., Gaggero, C., Medina, K., Fariña, L., Disegna, E., & Dellacassa, E. (2011). *Vitis vinifera* Tannat, chemical characterization and functional properties. Ten years of research. In R. Filip (ed.), *Multidisciplinary Approaches on Food Science and Nutrition for the XXI Century*. Research Signpost/ Transworld Res. Network. Kerala, India, 53-71.

[4] Da Silva, C., Zamperin, G., Ferrarini, A., Minio, A., Dal Molin, A., Venturini, L., Buson, G., Tononi, P., Avanzato, C., Zago, E., Boido, E., Dellacassa, E., Gaggero, C., Pezzotti, M., Carrau, F., & Delledonne, M. (2013). The high polyphenol content of grapevine

cultivar tannat berries is conferred primarily by genes that are not shared with the reference genome. *Plant Cell, 25*, 4777-4788.

[5] Carrau, F., Gaggero, C., & Aguilar, P. S. (2015). Yeast diversity and native vigor for flavor phenotypes. *Trends in biotechnology, 33*, 148-154.

[6] Medina, K., Martin, V., Boido, E., & Carrau, F. (2018). Yeast biotechnology for red winemaking. In A. Morata (Ed.), *Red Wine Technology* (pp. 69-83). Cambridge: Academic Press.

[7] Carrau, F., Boido, E., & Ramey, D. (2020). Yeasts for low input winemaking: microbial terroir and flavor differentiation. In G. M. Gaad & S. Sariaslani (Eds.), *Advances in Applied Microbiology*, (pp.89-121). Cambridge: Academic Press.

[8] González Techera, A., Jubany, S., Ponce de Leon, I., Boido, E. & Dellacassa, E. (2004). Molecular diversity within clones of cv. Tannat (*Vitis vinifera*). *Vitis, 43*(4), 179-185.

[9] Quideau, S., Deffieux, D., Douat-Casassus, C., & Pouységu, L. (2011). Plant polyphenols: chemical properties, biological activities, and synthesis. *Angewandte Chemie - International Edition, 50*(3), 586-621.

[10] Monagas, M., Bartolomé, B., & Gómez-Cordovés, C. (2005). Updated knowledge about the presence of phenolic compounds in wine. *Critical Reviews in Food Science and Nutrition, 45*, 85-118.

[11] Figueiredo-González, M., Martínez-Carballo, E., Cancho-Grande, B., Santiago, J. L., Martínez, M. C., & Simal-Gándara, J. (2012). Pattern recognition of three *Vitis vinifera* L. red grapes varieties based on anthocyanin and flavonol profiles, with correlations between their biosynthesis pathways. *Food Chemistry, 130*, 9-19.

[12] González-Neves, G., Gil, G., & Ferrer, M. (2002). Effect of different vineyard treatments on the phenolic contents in Tannat (*Vitis vinifera* L.) Grapes and their respective wines. *Food Science and Technology International, 8*, 315-321.

[13] González-Neves, G., Barreiro, L., Gil, G., Franco, J., Ferrer, M., Moutounet, M., & Carbonneau, A. (2004). Anthocyanic composition

of Tannat grapes from the south region of Uruguay. *Analytica Chimica Acta, 513*(1), 197-202.

[14] González-Neves, G., Charamelo, D., Balado, J., Barreiro, L., Bochicchio, R., Gatto, G., Gil, G., Tessore, A., Carbonneau, A., & Moutounet, M. (2004). Phenolic potential of Tannat, Cabernet-Sauvignon and Merlot grapes and their correspondence with wine composition. *Analytical Chimical Acta, 513*(1), 191-196.

[15] Boido, E., García-Marino, M., Dellacassa, E., Carrau, F., Rivas-Gonzalo, J. C. & Escribano-Bailón, M. T. (2011). *Characterisation and evolution of grape polyphenol profiles of Vitis vinifera L. cv. Tannat during ripening and vinification.*

[16] González-Neves, G., Gómez-Cordovés, C., & Barreiro, L. (2001). Anthocyanic composition of Tannat, Cabernet sauvignon and Merlot young red wines from uruguay. *Journal of Wine Research, 12*, 125-133

[17] Echeverry, C., Ferreira, M., Reyes-Parada, M., Abin-Carriquiry, J. A., Blasina, F., González-Neves, G., & Dajas, F. (2005). Changes in antioxidant capacity of Tannat red wines during early maturation. *Journal of Food Engineering, 69*, 147-154.

[18] Boido, E., Alcalde-Eon, C., Carrau, F., Dellacassa, E., & Rivas-Gonzalo, J. C. (2006). Aging effect on the pigment composition and color of *Vitis vinifera* L. cv. Tannat wines. Contribution of the main pigment families to wine color. *Journal of Agricultural and Food Chemistry, 54*, 6692-6704.

[19] Alcalde-Eon, C., Boido, E., Carrau, F., Dellacassa, E., & Rivas-Gonzalo, J. C. (2006). Pigment profiles in monovarietal wines produced in Uruguay. *American Journal of Enology and Viticulture, 57*, 449-459.

[20] Cerpa-Calderon, F. K., & Kennedy, J. (2008). Effect of berry crushing on skin and seed tannin extraction during fermentation and maceration. *American Journal of Enology and Viticulture. 59*, 350A-350A.

[21] Pinelo, M., Arnous, A., & Meyer, A. S. (2006). Upgrading of grape skins: Significance of plant cell-wall structural components and

extraction techniques for phenol release. *Trends in Food Science and Technology, 17,* 579-590.

[22] González-Neves, G., Gil, G., Favre, G., & Ferrer, M. (2012). Influence of grape composition and winemaking on the anthocyanin composition of red wines of Tannat. *International Journal of Food Science and Technology, 47,* 900-909.

[23] Amrani, K., & Glories, Y. (1995). Tanins et anthocyanes: Localisation dans le baie de raisin et mode d'extraction. *Reviews of French Oenology, 153,* 28-31. [Tannins and anthocyanins: Location in the grape berry and method of extraction]

[24] Sacchi, K. L., Bisson, L. F., & Adams, D. O. (2005). A review of the effect of winemaking techniques on phenolic extraction in red wines. *American Journal of Enology and Viticulture, 56,* 197-206.

[25] González-Neves, G., Favre, G., Piccardo, D., & Gil, G. (2016). Anthocyanin profile of young red wines of Tannat, Syrah and Merlot made using maceration enzymes and cold soak. *International Journal of Food Science and Technology, 51,* 260-267.

[26] Piccardo, D., González-Neves, G., Favre, G., Pascual, O., Canals, J. M., & Zamora, F. (2019). Impact of must replacement and hot pre-fermentative maceration on the color of uruguayan tannat red wines. *Fermentation, 5,* 80.

[27] Vasserot, Y., Caillet, S., & Maujean, A. (1997). Study of anthocyanin adsorption by yeast lees. Effect of some physicochemical parameters. *American Journal of Enology and Viticulture, 48,* 433-437.

[28] Morata, A., Gómez-Cordovés, M. C., Suberviola, J., Bartolomé, B., Colomo, B., & Suárez, J. A. (2003). Adsorption of anthocyanins by yeast cell walls during the fermentation of red wines. *Journal of Agricultural and Food Chemistry, 51,* 4084-4088.

[29] Medina, K., Boido, E., Dellacassa, E., & Carrau, F. (2005). Yeast interactions with anthocyanins during red wine fermentation. *American Journal of Enology and Viticulture, 56,* 104-109.

[30] Bakker, J., & Timberlake, C. F. (1997). Isolation, identification, and characterization of new color-stable anthocyanins occurring in some red wines. *Journal of Agricultural and Food Chemistry, 45,* 35-43.

[31] Medina, K., Boido, E., Dellacassa, E., & Carrau, F. (2018). Effects of non-*Saccharomyces* yeasts on color, anthocyanin, and anthocyanin-derived pigments of Tannat grapes during fermentation. *American Journal of Enology and Viticulture, 69,* 148-156.

[32] Medina, K., Boido, E., Fariña, L., Dellacassa, E., & Carrau, F. (2016). Non-*Saccharomyces* and *Saccharomyces* strains co-fermentation increases acetaldehyde accumulation: effect on anthocyanin-derived pigments in Tannat red wines. *Yeast, 33,* 339-343.

[33] Fariña, L., Carrau, F., Boido, E., Disegna, E., & Dellacassa, E. (2010). Carotenoid profile evolution in *Vitis vinifera* cv. Tannat grapes during ripening. *American Journal of Enology and Viticulture, 61,* 451-456.

[34] Brouillard, R. (1982). Chemical structure of anthocyanins. In P. Markakis (ed.), *Anthocyanins as Food Color.* London, Academic Press, 1-40.

[35] Downey, M. O., Harvey, J. S., & Robinson, S. P. (2003). Analysis of tannins in seeds and skins of Shiraz grapes throughout berry development. *Australian Journal of Grape and Wine Research, 9,* 15-27.

[36] Obreque-Slier, E., Peña-Neira, A., López-Solís, R., Zamora-Marín, F., Ricardo-Da Silva, J. M., & Laureano, O. (2010). Comparative study of the phenolic composition of seeds and skins from carménère and cabernet sauvignon grape varieties (*Vitis vinifera L.*) during ripening. *Journal of Agricultural and Food Chemistry, 58,* 3591-3599.

[37] Mattivi, F., Guzzon, R., Vrhovsek, U., Stefanini, M., & Velasco, R. (2006). Metabolite profiling of grape: Flavonols and anthocyanins. *Journal of Agricultural and Food Chemistry, 54,* 7692-7702.

[38] Guerrero, R. F., Liazid, A., Palma, M., Puertas, B., González-Barrio, R., Gil-Izquierdo, A., García-Barroso, C., & Cantos-Villar, E.

(2009). Phenolic characterisation of red grapes autochthonous to Andalusia. *Food Chemistry, 112*, 949-955.

[39] Tian, R. R., Li, G., Wan, S. B., Pan, Q. H., Zhan, J. C., Li, J. M., Zhang, Q. H., & Huang, W. D. (2009). Comparative study of 11 phenolic acids and five flavan-3-ols in cv. vidal: Impact of natural icewine making versus concentration technology. *Australian Journal of Grape and Wine Research, 15*, 216-222.

[40] Peynaud, E. (1987). *The taste of wine: The art and science of wine appreciation* London: Macdonald Orbis.

[41] Lyman, B. J., & Green, B. G. (1990). Oral astringency: effects of repeated exposure and interactions with sweeteners. *Chemical Senses, 15*, 151-164.

[42] Guinard, J., Pangborn, R., & Lewis, M. (1986). The time-course of astringency in wine upon repeated ingestion. *American Journal of Enology and Viticulture, 37*, 184-189.

[43] Vidal, L., Antúnez, L., Giménez, A., Medina, K., Boido, E., & Ares, G. (2017). Sensory characterization of the astringency of commercial Uruguayan Tannat wines. *Food Research International, 102*, 425-434.

[44] Vidal, L., Antúnez, L., Giménez, A., Medina, K., Boido, E., & Ares, G. (2016). Dynamic characterization of red wine astringency: Case study with Uruguayan Tannat wines. *Food Research International, 82*, 128-135.

[45] Vidal, L., Antúnez, L., Rodríguez-Haralambides, A., Giménez, A., Medina, K., Boido, E., & Ares, G. (2018). Relationship between astringency and phenolic composition of commercial Uruguayan Tannat wines: Application of boosted regression trees. *Food Research International, 112*, 25-37.

[46] Renaud, S., & de Lorgeril, M. (1992). Wine, alcohol, platelets, and the French paradox for coronary heart disease. *The Lancet, 339*, 1523-1526.

[47] Lin, J. K., & Weng, M. S. (2006). Flavonoids as nutraceuticals. In E. Grotewold (Ed.), *The Science of Flavonoids* (pp. 213-238). New York: Springer.

[48] Cory, H., Passarelli, S., Szeto, J., Tamez, M., & Mattei, J. (2018). The Role of Polyphenols in Human Health and Food Systems: A Mini-Review. *Frontiers in Nutrition, 5*, 1-9.

[49] Khan, N., Adhami, V. M., & Mukhtar, H. (2010). Apoptosis by dietary agents for prevention and treatment of prostate cancer. *Endocrine-Related Cancer, 17*, 39-52.

[50] Fitzpatrick, D. F., Hirschfield, S. L., & Coffey, R. G. (1993). Endothelium-dependent vasorelaxing activity of wine and other grape products. *American Journal of Physiology - Heart and Circulatory Physiology, 265* (2), 774-778.

[51] Hashimoto, M., Kim, S., Eto, M., Iijima, K., Ako, J., Yoshizumi, M., Akishita, M., Kondo, K., Itakura, H., Hosoda, K., Toba, K., & Ouchi, Y. (2001). Effect of acute intake of red wine on flow-mediated vasodilatation of the brachial artery. *American Journal of Cardiology, 88*, 1457-1460.

[52] Corder, R., Mullen, W., Khan, N. Q., Marks, S. C., Wood, E. G., Carrier, M. J., & Crozier, A. (2006). Oenology: Red wine procyanidins and vascular health. *Nature, 444*, 566.

[53] Auvray, M., & Spence, C. (2008). The multisensory perception of flavor. *Consciousness and Cognition, 17*, 1016-1031.

[54] Chang, S. S. (1989). Food flavors. *Food technology, 43*, 99-106.

[55] Van Gemert, L. J. (2011). *Flavour thresholds: Compilations of flavour threshold values in water and other media*. The Netherlands: Oliemans Punter.

[56] Versini, G., Dellacassa, E., Carlin, S., Fedrizzi, B., & Magno, F. (2008). Analysis of aroma compounds in wine. In R. Flamini (Ed.), *Hyphenated techniques in grape and wine chemistry* (pp. 173-225). New Jersey: John Wiley & Sons.

[57] Versini, G., Carlin, S., Dalla Serra, A., Nicolini, G., & Rapp, A. (2002). Formation of 1,1,6-trimethyl-1,2-dihydronaphthalene and other norisoprenoids in wine: Considerations on the kinetics. In P. Winterhalter, & R. L. Rouseff (Eds.), *Carotenoid Derivedc Aroma Compounds* (pp. 285-299). Washington: American Chemical Society.

[58] Boido, E., Fariña, L., Carrau, F., Dellacassa, E., & Cozzolino, D. (2013). Characterization of glycosylated aroma compounds in Tannat grapes and feasibility of the near infrared spectroscopy application for their prediction. *Food Analytical Methods, 6*, 100-111.

[59] Fariña, L. (2008). *Caracterización del Perfil aromático de vinos Tannat y su evolución durante la crianza* (PhD Thesis). Universidad de la Republica, Montevideo, Uruguay. [*Characterization of the aromatic profile of Tannat wines and its evolution during aging*]

[60] Boido, E., Fariña, L., Capra, A., Medina, K., Coniberti, A., Disegna, E., Dellacassa, E., & Carrau, F. (2005). ¿Es posible predecir la calidad de la fruta para la vinificación mediante parámetros de fácil determinación en bodega? Estudio para la variedad Tannat. *Annals of the Latin American Congress of Viticulture and Enology*, Bento Gonçalves, RS, Brasil, 10. [Is it possible to predict the quality of the fruit for winemaking using parameters that can be easily determined in the winery? Study for the Tannat variety. *Annals of the Latin American Congress of Viticulture and Enology*]

[61] Félix, E. (2003). *Efecto de diferentes intensidades y momentos de deshojado en Vitis vinifera L. cv. Tannat sobre parametros cualicuantitativos de la uva y el vino* (MSc Thesis). Universidad de la República, Montevideo, Uruguay. [*Effect of different intensities and defoliation moments in Vitis vinifera L. cv. Tannat on qualiquantitative parameters of grapes and wine*]

[62] Gómez-Plaza, E., Gil-Muñoz, R., & Martínez-Cutillas, A. (2000). Multivariate classification of wines from seven clones of Monastrell grapes. *Journal of the Science of Food and Agriculture, 80*, 497-501.

[63] Disegna, E., Coniberti, A., & Ferrari, V. (2014). Clones de Tannat en Uruguay. *Revista del Instituto Nacional de Investigación Agropecuaria, 109*, 1-6. [Tannat clones in Uruguay. *Magazine of the National Institute of Agricultural Research*]

[64] Fariña, L., Coniberti, A., Passarino, E., Zapater, M., Acuña, O., Boido, E., Carrau, F., Disegna, E., & Dellacasa, E. (2018). Characterization of aromatic precursors in commercial *Vitis vinifera*

cv Tannat clones present in Uruguay. *Annals of the World Congress of Vine and Wine.* Punta del Este, Uruguay, 41.

[65] Gámbaro, A., Boido, E., Zlotejablko, A., Medina, K., Lloret, A., Dellacassa, E., & Carrau, F. (2001). Effect of malolactic fermentation on the aroma properties of Tannat wine. *Australian Journal of Grape and Wine Research, 7,* 27-32.

[66] Boido, E., Lloret, A., Medina, K., Fariña, L., Carrau, F., Versini, G., & Dellacassa, E. (2003). Aroma composition of V*itis vinifera* Cv. Tannat: the typical red wine from Uruguay. *Journal of Agricultural and Food Chemistry, 51,* 5408-5413.

[67] Boido, E., Lloret, A., Medina, K., Carrau, F., & Dellacassa, E. (2002). Effect of β-glycosidase activity of Oenococcus oeni on the glycosylated flavor precursors of Tannat wine during malolactic fermentation. *Journal of Agricultural and Food Chemistry, 50,* 2344–2349.

[68] Pérez, G., Fariña, L., Barquet, M., Boido, E., Gaggero, C., Dellacassa, E., & Carrau, F. (2011). A quick screening method to identify β-glucosidase activity in native wine yeast strains: Application of Esculin Glycerol Agar (EGA) medium. *World Journal of Microbiology and Biotechnology, 27,* 47-55.

[69] Hu, K., Qin, Y., Tao, Y-S., Zhu, X-L., Peng, C-T., & Ullah, N. (2016). Potential of Glycosidase from Non- *Saccharomyces* Isolates for Enhancement of Wine Aroma. *Journal of Food Science, 81,* M935-M943.

[70] López, S., Mateo, J., & Maicas, S. (2015). Screening of *Hanseniaspora* strains for the production of enzymes with potential interest for winemaking. *Fermentation, 2,* 1-16.

[71] Hernández-Orte, P., Cersosimo, M., Loscos, N., Cacho, J., Garcia-Moruno, E., & Ferreira, V. (2008). The development of varietal aroma from non-floral grapes by yeasts of different genera. *Food Chemistry, 107,* 1064-1077.

[72] Cordero Otero, R. R., Ubeda Iranzo, J. F., Briones-Perez, A. I., Potgieter, N., Villena, M. A., Pretorius, I. S., & van Rensburg, P. (2003). Characterization of the β-glucosidase activity produced by

enological strains of non-*Saccharomyces* Yeasts. *Journal of Food Science, 68*, 2564-2569.

[73] Fariña, L., Villar, V., Ares, G., Carrau, F., Dellacassa, E., & Boido, E. (2015). Volatile composition and aroma profile of Uruguayan Tannat wines. *Food Research International, 69*, 244-255.

[74] Razungles, A., Gunata, Z., Pinatel, S., Baumes, R., & Bayonove, C. (1993). Etude quantitative de composés terpéniques, norisoprénoïdes et de leurs précurseurs dans diverses variétés de raisins. *Sciences des aliments, 13*, 59-72. [Quantitative study of terpene compounds, norisoprenoids and their precursors in various varieties of grapes. *Food science*]

[75] Escudero, A., Campo, E., Fariña, L., Cacho, J., & Ferreira, V. (2007). Analytical characterization of the aroma of five premium red wines. Insights into the role of odor families and the concept of fruitiness of wines. *Journal of Agricultural and Food Chemistry, 55*, 4501-4510.

[76] Molina, A. M., Guadalupe, V., Varela, C., Swiegers, J. H., Pretorius, I. S., & Agosin, E. (2009). Differential synthesis of fermentative aroma compounds of two related commercial wine yeast strains. *Food Chemistry, 117*, 189-195.

[77] Hernández-Orte, P., Cacho, J. F., & Ferreira, V. (2002). Relationship between varietal amino acid profile of grapes and wine aromatic composition. Experiments with model solutions and chemometric study. *Journal of Agricultural and Food Chemistry, 50*, 2891-2899.

[78] Medina, K., Boido, E., Fariña, L., Dellacassa, E., & Carrau, F. (2018). Impact on Tannat wines aroma produced by different yeast using three vinification systems. *Annals of the World Congress of Vine and Wine.* Punta del Este, Uruguay, 41.

[79] Martin, V., Valera, M., Medina, K., Boido, E., & Carrau, F. (2018). Oenological Impact of the Hanseniaspora/Kloeckera Yeast Genus on Wines-A Review. *Fermentation, 4*, 76.

[80] Boido, E., Medina, K., Fariña, L., Carrau, F., Versini, G., & Dellacassa, E. (2009). The effect of bacterial strain and aging on the secondary volatile metabolites produced during malolactic

fermentation of Tannat red wine. *Journal of Agricultural and Food Chemistry, 57,* 6271-6278.

[81] De La Presa-Owens, C., & Noble, A. C. (1995). Descriptive Analysis of Three White Wine Varieties from Penedès. *American Journal of Enology and Viticulture, 46,* 5-9.

[82] Heymann, H., Noble, A. C., & Boulton, R. B. (1986). Analysis of methoxypyrazines in wines. 1. Development of a quantitative procedure. *Journal of Agricultural and Food Chemistry, 34,* 268-271.

[83] Noble, A. C., Williams, A. A., & Langron, S. P. (1984). Descriptive analysis and quality ratings of 1976 wines from four Bordeaux communes. *Journal of the Science of Food and Agriculture, 35,* 88-98.

[84] Hootman, R., & Stone, H. (2008). Quantitative Descriptive Analysis (QDA). In R. Hootman (Ed.), *Manual on Descriptive Analysis Testing for Sensory Evaluation* ASTM International, (pp. 15-21). West Conshohocken: ASTM Inernational.

[85] Gámbaro, A., Boido, E., Zlotejablko, A., Medina, K., Lloret, A., Dellacassa, E., & Carrau, F. (2001). Effect of malolactic fermentation on the aroma properties of Tannat wine. *Australian Journal of Grape and Wine Research, 7,* 27-32.

[86] Ares, G., & Varela, P. (2014). Comparison of novel methodologies for sensory characterization. In P. Varela, & G. Ares (Eds.), *Novel techniques in sensory characterization and consumer profiling,* (pp. 378-403). Boca Raton: CRC Press.

[87] Prescott, J. (1999). Flavour as a psychological construct: Implications for perceiving and measuring the sensory qualities of foods. *Food Quality and Preference, 10,* 349-356.

[88] Varela, P., & Ares, G. (2012). Sensory profiling, the blurred line between sensory and consumer science. A review of novel methods for product characterization. *Food Research International, 48,* 893-908.

In: Fermented and Distilled
Editors: M. B. M. de Castilhos et al.
ISBN: 978-1-53618-985-8
© 2021 Nova Science Publishers, Inc.

Chapter 7

SYRAH (*VITIS VINIFERA* L.) WINES IN BRAZIL

Juliane Barreto de Oliveira[*] *and Giuliano Elias Pereira*[†]

Brazilian Agricultural Research Corporation, Embrapa Grape & Wine,
Bento Gonçalves, Rio Grande do Sul, Brazil

ABSTRACT

The grapevine Syrah is originated from France and ir grows in several winegrowing regions around the world. It is a grape variety used successfully worldwide, producing wines with high variability of styles, with complexity and providing numerous aromas and flavors, in young and/or aged/guard wines. The most valued assets of Syrah is the capacity to adapt to different climates and soils due to its plasticity, being commercially used over many years ago in several countries. In this chapter, we show the main characteristics of Syrah variety worldwide, with researches carried out in different winegrowing zones in Brazil, in traditional sites, and also in new viticultural regions. The results show the variation of the physicochemical composition of the grapes and wines, as

---

[*] Corresponding Author's E-mail: juliane.barreto@hotmail.com.
[†] Corresponding Author's E-mail: giuliano.pereira@embrapa.br.

well as the sensory profiles of wines mainly influenced by different climates, rootstocks, and winemaking protocols.

Keywords: *Vitis vinifera* L., grape, wine, chemical composition, geographical conditions, phenolic compounds, sensory profile

INTRODUCTION

Since its origin from Asia, grape, wine, and its derived products have become one of the most important socio-economic activities in the world [1]. The global grape production in 2018 was around 77.8 million tons, showing modest increases in the last decade [2]. Approximately 57% of the world production was destined for processing, which 92% for wines, and 8% for grape juices and musts, 36% for fresh consumption, and 7% for dried grape [2]. The destination of grape production varies greatly depending on the country.

Syrah is a noble wine grape, planted in all five continents in the world, and is attracting more and more interest to the consumers and the productive sector, mainly due to its capacity to adapt to different *"terroirs"*, presenting a high diversity of products and many styles of wines, with undeniable quality, in young and/or guard wines [3]. Over the years, studies have been carried out with the variety, in different producing regions of the world, in order to better understand its behavior to different environmental factors, such as climate, soil, and others, as well as to characterize the influence of natural and/or induced parameters (by a human factor) on grape and wine composition [4-12].

Brazil is the only country around the world where it is possible to produce three kinds of wines, mainly due to the climate conditions (the geography), and the vine managements [13]. The first winegrowing condition is the traditional viticulture, commercially used in temperate and subtropical climates in Southern and Southeastern Brazil, where vines are pruned and harvested just once a year, like in the majority countries of the world, in both Hemispheres ("traditional wines"). The second kind of viticulture is located in the São Francisco Valley, in Northeastern Brazil, in a tropical semi-arid climate, where "tropical wines" have been produced since 1985, pruning and harvesting the same vine twice a year [14]. Adopting the scheduling of different plots and vineyards, with high temperatures, solar radiation, water availability for irrigation, and the use of phytoregulators, it is possible to prune and harvest every day of the year [15]. In this condition, wine typicality varies due to intra-annual climate variability, resulting in very different wines, depending on the harvest date [11]. Finally, a third winegrowing zone started in 2004 in Southeastern Brazil, between 600-1.200 m of elevation above sea level, where the vines are pruned twice and grapes are harvested once per year, producing the called "winter wines". In this area, vine management requires also the use of phytoregulators to allow two vegetative cycles (formation and production).

Syrah wines have being produced in all of these three winegrowing regions of Brazil, but very few in traditional zones (South). In this chapter, we will present the main characteristics of the grapevine in two regions, in the Northeast, where tropical and winter wines are made, and in the Southeast, where wineries also make winter wines.

Origin of the Variety

Although there is a long-documented history about the production of Syrah (*Vitis vinifera* L.) vineyards in the Rhone region, in southeastern France, it was not certain whether this variety has French origin. But in 1998, a study conducted by the research group of the Department of Viticulture and Enology at the University of California (Davis) employed DNA typing and extensive reference material of grapes from the viticultural research station in Montpellier, France, to conclude that Syrah is descended from the Dureza and Mondeuse Blanche varieties [16,17]. Over time, it has expanded to many other countries, being today one of the most planted red varieties in the world [18].

Figure 1. Syrah grapes (*Vitis vinifera* L.). Source: Oliveira, J.B.

Nomenclature

Regarding the nomenclature, this variety is called Syrah in its country of origin (France), as well as in the rest of Europe, Argentina, Brazil, Chile, New Zealand, Uruguay, and in the United States. The name Shiraz has became popular for this grape variety in Australia, where it has long been established as the most widely cultivated red variety. The name Shiraz is also commonly used in South Africa, the United States, and Canada. This grape is also known under other synonyms, used in many parts of the world, including Antourenein Noir, Balsamina, Candive, Entournerein, Hignin Noir, Marsanne Noir, Schiras, Sirac, Syra, Syrac, Serine, and Sereine [19].

Ampelography Characteristics

The leaves are medium-sized, with five lobes, and slightly sinuated. The berries are small, oval, blue-black color, with smooth but strong skin, covered with abundant pruin (Figure 2). The mesocarp is juicy, with a pleasant taste, without flavor. There exist seven certified Syrah clones commercially available, identified as 470, 471, 524, 747, 1140, 1141, and 1188 [20].

Figure 2. Syrah (*Vitis vinífera* L.) - Leaf and cluster. Source: Wikimedia Commons.

In traditional viticultural regions in temperate climates, the period of vegetation of the variety is long, ranging between 190-210 days [20, 21], the maturation is late. In a tropical semi-arid climate in Brazil, such as in the São Francisco Valley, its cycle is very short and varies from 110 to 150 days, depending on the grape destination. In this region, the great part of the Syrah grapes are used to make young red wines, but also it is used for white and rosé sparkling wines, as well as a very little part destined to aging wines. For this purpose, the better grapes are those harvested between May and August [11, 15].

Production of Syrah Grapes and Wines in the World

Data reported that Syrah is planted in several countries around the world (Figure3), in a total area of 190.000 ha, and France (64.000 ha), Australia (40.000 ha), Spain (20.000 ha), and Argentine (13.000 ha) are the main producers [18]. Syrah is also grown in other countries, such as Germany, Mexico, New Zealand, Romania, Switzerland, Brazil, among others.

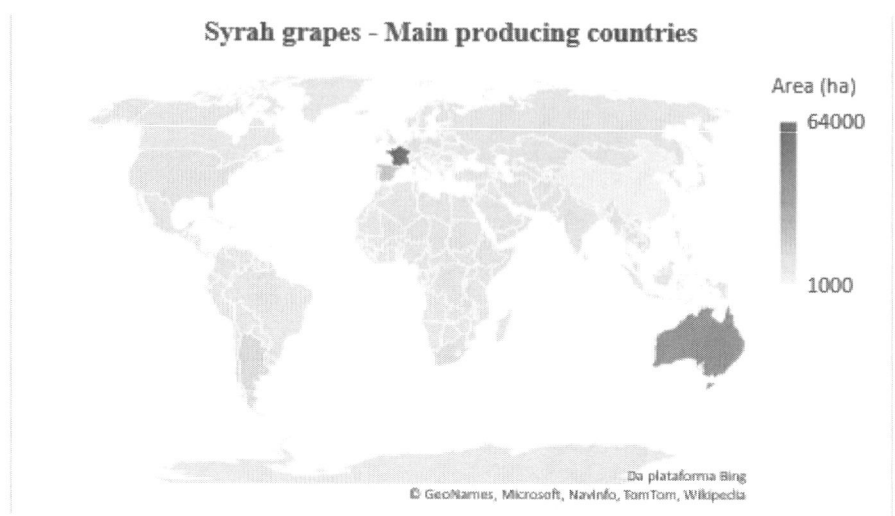

Figure 3. Important regions producing the Shiraz variety (*Vitis vinifera* L.) in the world. Source: OIV, 2019 [2].

Wine production in 2018 was estimated at 292 mhL, and the EU and the Americas were the largest producers in Figure 4 [2]. Some factors, including grape variety, ripeness, environmental factors (such as climate and soil), and technological procedures used during winemaking, can qualitatively and quantitatively affect the chemical and phenolic composition of the grape and wine [22]. Therefore, the chemical and sensory profile of wines of the same variety, grown in different regions, can be very different.

Syrah is used to make varietal red wines (young or aged), as well as rosé and sparkling wines [11, 15]. It is widely used in blends with other varieties such as Cabernet Sauvignon, Merlot, Touriga Nacional, Tempranillo, among other varieties.

The grape and wine composition vary strongly according to the "*terroir*" in every winegrowing regions worldwide. *Terroir* is the effect of climate, soil, and human factor on wine composition and typicality [23, 24]. The grape composition is dependent on several factors, such as climate, soil, rootstock, the clone of the variety, yield, irrigation, nutrition, harvest date, and others [11, 25]. The main compounds responsible for wine composition desirable in grapes at harvest are the sugars, which will be converted in alcohol content; organic acids (tartaric, malic, lactic, citric, and others), responsible to total acidity, balance, freshness, the stability of the wines; nitrogen content, such as amino acids and others, which will be used by yeasts as energy to convert sugars; primary aromas and some precursors, and phenolic compounds, representing color, structure/body, and also the wine balance [26-28].

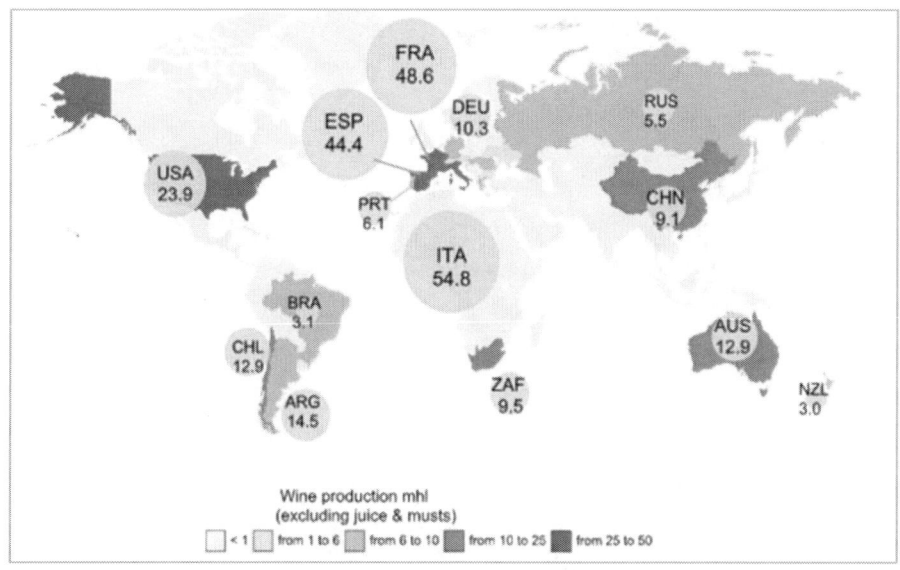

Figure 4. Wine production in different regions in the world. Source: OIV, 2019 [2].

The wine quality depends on the grape composition, and the final concentration and the balance between all compounds can vary according to the kind of wine elaborated. For example, sparkling wines are elaborated with grapes presenting high acidity, and few phenolic compounds, to avoid bitterness in the taste. White wines can have different concentrations, but the composition required is a balance between alcohol degree and total acidity, avoiding also phenolic extraction from skin and seeds [27]. For red wines, there are a lot of products, among young and guard/aged wines, but the balance which every winery, producer, and consumer look at is the balance between alcohol degree, total acidity, and phenolic composition (tannins) in the taste/mouth.

Among the wide chemical composition of wines, phenolic compounds are a group of natural substances of secondary metabolites produced by grapevines that play an important role in wine quality [26, 27]. Most of these compounds are derived from those contained in grapes and extracted from skins, seeds, and pulps during the initial stages of winemaking. In recent years, the phenolic and aromatic characterization of wines has become crucial and important to verify the typicality and identity of a specific region. Therefore, several studies have been carried out over the years. In Table 1, there is some data of wine composition, mainly related to phenolic compound concentrations in different traditional regions.

The major phenolic compounds in Syrah grape, and consequently in the wine, include the anthocyanins red pigments and condensed tannins, also named as proanthocyanidins [11]. The latter is present in all berry tissues but the structure of seed proanthocyanidins is different form as compared to those present in skins. The phenolic composition of Syrah grapes depends primarily on the berry development since proanthocyanidins are biosynthesized at the green stage, while anthocyanins accumulate after veraison, usually reaching a peak before harvest, then starting to decline.

Wine phenolic composition includes compounds extracted from the grape during maceration process, and products formed through enzymatic and chemical reactions [29]. They are relevant for the visual and taste profile (in sensory analysis), as well for contributing to the aging process.

Tannins are very complex molecules, with different sizes and structures, which play an important role in the perception of astringency and body in red wines. As well as anthocyanins, tannin extraction in the wine depends on several factors, including the variety, the maturation level at harvest, also the extractability (of the seeds and skins) during the winemaking process. Studies have shown that Syrah wines, produced in different countries, tend to have high concentrations of tannins (Table 1) when compared to other varieties, such as Pinot Noir [37], Trincadeira, and Cabernet Sauvignon [38].

Table 1. Phenolic composition of Syrah red wines in different countries

Compound	Concentration	Country	Reference
Total Phenols (mg.L^{-1})			
	1954	Greece	[5]
	1677 - 1835	Chile	[6]
	1181 - 3289	Australia	[4]
	1422	Argentina	[7]
Total monomeric anthocyanins (mg.L^{-1})			
	140	Spain	[30]
	22- 259	Australia	[31]
	207 - 416	Australia	[32]
	394 - 652	Chile	[33]
Tannins (mg.L^{-1})			
	930 - 1650	Chile	[6]
	264 - 298	Spain	[33]
	949 - 1478	Australia	[34]
Trans-resveratrol (mg.L^{-1})			
	1.1 - 2.4	Chile	[6]
	0.5 - 2.7	Argentina	[7]
	1.6 - 2.0	Uruguay	[35]
	0.64	Italy	[36]
	2.0	California	[4]

Production of Syrah Grapes and Wines in Brazil

In Brazil, the total cultivated area with grapevines is 75.951 ha in different regions, and the South (73.35%), Southeast (11.48%), and Northeast (14.87%) are the main producers, with a grape production destined for processing of 818.287 ton [39]. Actually, in Brazil there are three different winegrowing zones producing different kinds of wines, according to climate conditions and grapevine managements (Figure 5). Three kinds of wines were characterized, described as traditional wines in the South and Southeast regions, tropical wines in the Northeast, and winter wines in the Southeast, Centerwest and Northeast, but at high elevations, between 600 m to 1,200 m above sea level [40].

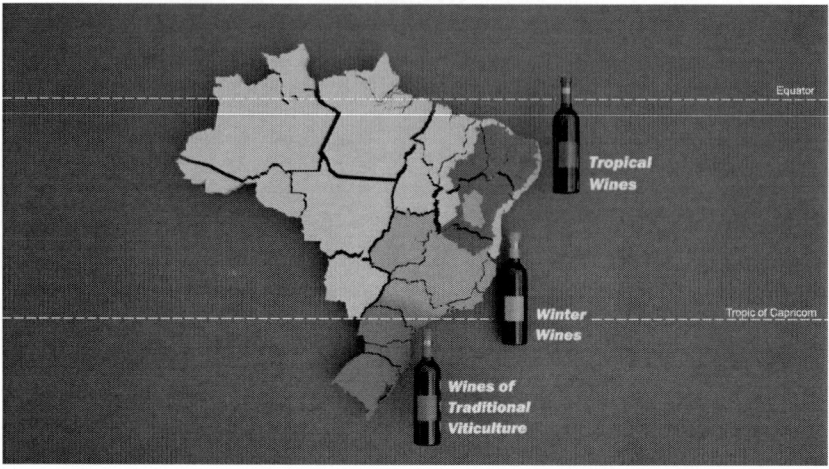

Figure 5. Traditional, tropical and winter wines produced in Brazil. Source: Pereira et al. 2020 [40].

Traditional Syrah Wines in Brazil

The Syrah variety is cultivated in most parts of Brazil. In the Serra Gaúcha, located in the South region in a temperate and sub-tropical climate, the most known winegrowing zone of the country, representing around 60% of total vineyards in Brazil. In this traditional viticulture, Syrah is present in only two hectares, producing two high-quality red wines, one of them young, and other one a guard/aged. In other parts of Brazil, such as in the tropical semi-arid climate of the Northeast (São Francisco Valley) and the sub-tropical climate in the Southeast, as well as in the tropical of altitude climate of the North-East (Chapada Diamantina, Bahia State), it has presented an excellent performance, and it is cultivated in medium scale.

Tropical Syrah Wines in Brazil

Under the climate conditions of the Sub-middle of the São Francisco Valley (tropical viticulture), Syrah is cultivated in around 60 hectares of vineyards, being an early cultivar. The berry maturation starts from 47 to 55 days after fruit set, corresponding to 75-85 days after pruning, depending on the period of the year [11, 12]. The cycle duration range from 110 to 150 days, depending on the wine elaborated. It is currently the main grape variety used in the São Francisco Valley for red wines, but it is also successfully used to make white and rosé sparkling wines, in blends with other varieties, such as Tempranillo, Grenache, and others [15]. It is interesting to highlight the high versatility and potential of the variety to be used in different products. Also, if a specific plot is managed to elaborate red wines, and if the winery has a demand and a lack of a sparkling wine stored, they do not hesitate in harvesting that plot with Syrah, primarily destined to make red wine, and elaborate the sparkling. This is the main characteristic of the region, where the focus of the enterprises is to make sparkling wines firstly, followed by young red wines, then white and few aged red wines [15, 41].

In these conditions, some research was carried out in the last ten years, in which Syrah grapes were characterized and compared according to different harvest dates in the same year, as well as different rootstocks (Table 2) [42]. It is interesting to note that grape composition varied according to the harvest date. Anthocyanins in the grapes harvested in the first semester presented the highest concentrations as compared to the grapes harvested from the same vines in the second semester. The effect of the rootstock was also significant for anthocyanins and varied according to different compounds [42].

The seeds of Syrah grafted onto 1103P rootstock had higher values of total condensed tannins than the seeds of Syrah grafted onto IAC 313 rootstock. It demonstrated that the tannin profile in seeds might be related to variety and rootstock interaction, whereas the concentrations should be linked to the climatic conditions of the region. There is a tendency for higher concentration of compounds in the grapes of the first harvest season of the year (between April to June) [42].

Table 2. Phenolic composition of Syrah grapes in different seasons and rootstocks, in a tropical semi-arid climate condition of Brazil [42]

Variety vs. Rootstock	Syrah – 1103P			
Experimental assays	1st Year		2nd Year	
Harvest season	First	Second	First	Second
Monomeric anthocyanins (mg g^{-1})				
Delphinidin 3-O-glucoside	1.6^c ±0.3	0.0^e ±0.0	4.7^b ±0.1	0.6^d ±0.0
Cyanidin 3-O-glucoside	0.3^{de} ±0.0	2.2^{ab} ±0.2	0.7^c ±0.1	2.3^a ±0.0
Peonidin 3-O-glucoside	4.0^c ±0.2	1.2^d ±0.1	4.7^a ±0.2	4.1^{bc} ±0.0
Petunidin 3-O-glucoside	1.1^{de} ±0.0	1.5^d ±0.0	3.3^b ±0.0	1.1^{de} ±0.1
Malvidin 3-O-glucoside	15.8^{de} ±0.6	13.8^f ±0.7	27.7^b ±0.0	16.6^d ±0.5
Peonidin 3-O-acetylglucoside	0.3^e ±0.1	1.2^{bc} ±0.1	0.3^e ±0.0	0.8^d ±0.0
Petunidin 3-O-acetylglucoside	1.3^{cd} ±0.1	1.6^{cd} ±0.1	2.6^b ±0.1	1.1^e ±0.0
Cyanidin 3-O-acetylglucoside	0.2^{bc} ±0.0	0.0^d ±0.0	0.4^{ab} ±0.0	0.0^d ±0.0
Delphinidin 3-O-acetylglucoside	0.7^c ±0.1	0.0^d ±0.0	1.2^b ±0.1	0.3^{cd} ±0.0
Malvidin 3-O-acetylglucoside	6.5^d ±0.3	6.6^{cd} ±0.2	2.3^f ±0.0	17.0^a ±0.2
Peonidin 3-O-coumarylglucoside	0.6^{de} ±0.3	1.3^c ±0.1	2.0^{ab} ±0.1	0.7^d ±0.0
Petunidin 3-O-coumarylglucoside	0.7^d ±0.0	1.6^a ±0.2	1.3^b ±0.0	1.6^a ±0.2
Delphinidin 3-O-coumarylglucoside	3.1^c ±0.4	1.1^e ±0.0	2.8^{cd} ±0.0	3.3^c ±0.0
Malvidin 3-O-coumarylglucoside	7.1^{ab} ±0.7	4.2^d ±1.0	7.0^b ±0.0	2.5^f ±0.2
Total monomeric anthocyanins	43.3^d ±1.0	36.3^e ±0.9	61.0^b ±0.2	52.0^c ±0.4
Condensed tannins in seeds				
Monomeric (mg g^{-1})	2.8^a ±0.0	2.1^{bc} ±0.0	1.0^d ±0.0	2.3^b ±0.5
Oligomeric (mg g^{-1})	9.8^b ±0.9	3.9^f ±0.0	3.6^g ±0.0	4.1^f ±0.0
Polymeric (mg g^{-1})	33.8^b ±0.5	20.2^e ±0.0	23.6^{de} ±0.5	19.4^f ±0.1
Total tannins in seeds (mg g^{-1})	46.4^b ±1.0	26.2^g ±0.5	28.2^f ±0.5	25.8^g ±0.5
Condensed tannins in skins				
Monomeric (mg g^{-1})	0.1^{ab} ±0.0	0.0^b ±0.0	0.1^{ab} ±0.0	0.1^{ab} ±0.0
Oligomeric (mg g^{-1})	0.2^c ±0.1	0.5^b ±0.0	0.6^b ±0.1	0.3^c ±0.0
Polymeric (mg g^{-1})	1.5^d ±0.2	1.7^{cd} ±0.0	2.8^a ±0.3	1.8^c ±0.0
Total tannins in skins (mg g^{-1})	1.8^e ±0.0	2.2^d ±0.5	3.5^a ±0.3	2.2^d ±0.2
Monomeric anthocyanins (mg g^{-1})				
Delphinidin 3-O-glucoside	1.6^c ±0.1	0.0^e ±0.0	5.3^a ±0.2	0.4^{de} ±0.0
Cyanidin 3-O-glucoside	0.2^e ±0.0	0.7^c ±0.0	0.7^c ±0.0	0.5^{cd} ±0.1
Peonidin 3-O-glucoside	4.4^b ±0.1	0.0^e ±0.0	4.6^a ±0.3	1.5^d ±0.3
Petunidin 3-O-glucoside	2.6^c ±0.3	0.7^e ±0.0	3.9^a ±0.1	2.4^c ±0.3
Malvidin 3-O-glucoside	26.1^c ±0.3	14.3^e ±0.1	33.6^a ±0.5	16.5^d ±0.8
Peonidin 3-O-acetylglucoside	0.8^d ±0.1	0.3^e ±0.0	1.4^b ±0.1	1.8^a ±0.0
Petunidin 3-O-acetylglucoside	1.1^e ±0.0	0.6^{ef} ±0.0	5.4^a ±0.2	0.4^f ±0.1
Cyanidin 3-O-acetylglucoside	0.0^d ±0.0	0.0^d ±0.0	0.3^{ab} ±0.0	0.1^c ±0.0
Delphinidin 3-O-acetylglucoside	0.4^{cd} ±0.0	0.0^d ±0.0	1.3^{ab} ±0.1	1.4^{ab} ±0.1

Variety vs. Rootstock	Syrah – 1103P			
Experimental assays	1st Year		2nd Year	
Harvest season	First	Second	First	Second
Malvidin 3-O-acetylglucoside	7.0^c ±0.4	1.8^g ±0.0	3.4^e ±0.0	8.4^b ±1.4
Peonidin 3-O-coumarylglucoside	0.8^{cd} ±0.0	0.5^e ±0.0	0.4^e ±0.0	2.1^a ±0.1
Petunidin 3-O-coumarylglucoside	0.2^f ±0.0	0.9^{cd} ±0.0	0.6^{de} ±0.0	0.7^d ±0.0
Delphinidin 3-O-coumarylglucoside	6.6^b ±0.2	0.0^f ±0.0	15.9^a ±0.4	0.7^{ef} ±0.3
Malvidin 3-O-coumarylglucoside	4.9^c ±0.1	1.8^g ±0.1	7.4^a ±0.0	3.4^e ±0.3
Total monomeric anthocyanins	56.7^{bc} ±0.9	21.6^f ±0.1	84.2^a ±1.3	40.3^d ±1.6
Variety vs. Rootstock	Syrah – 1103P			
Experimental assays	1st Year		Experimental assays	
Harvest season	First	Harvest	First	Harvest
Condensed tannins in seeds				
Monomeric (mg g^{-1})	2.7^a ±0.1	2.3^b ±0.3	1.1^d ±0.0	1.8^c ±0.0
Oligomeric (mg g^{-1})	12.1^a ±0.7	7.5^c ±0.2	5.6^e ±0.2	6.3^d ±0.0
Polymeric (mg g^{-1})	32.3^c ±0.5	25.1^d ±0.2	33.4^b ±0.3	34.6^a ±0.5
Total tannins in seeds (mg g^{-1})	47.1^a ±0.7	34.9^e ±0.3	40.1^d ±0.0	42.7^c ±0.8
Condensed tannins in skins				
Monomeric (mg g^{-1})	0.1^{ab} ±0.0	0.0^b ±0.0	0.2^a ±0.0	0.1^{ab} ±0.0
Oligomeric (mg g^{-1})	0.6^b ±0.0	0.2^c ±0.0	0.8^a ±0.0	0.3^c ±0.0
Polymeric (mg g^{-1})	1.8^c ±0.1	1.4^d ±0.0	2.0^{bc} ±0.0	2.2^{bc} ±0.4
Total tannins in skins (mg g^{-1})	2.5^c ±0.1	1.6^d ±0.0	3.0^b ±0.0	2.6^c ±0.3

Means followed by the same letter within rows did not differ by Tukey's test at 5% (p < 0.05). Legend: * mg g^{-1} of grapes; I Harvest (first harvest season, years 2016 and 2017); II Harvest (second harvest season, years 2014 and 2016).

Winter Syrah Wines in Brazil

The grapevine Syrah was introduced in two states to produce winter wines, firstly in Minas Gerais, in 2004, and then in other states [43-46]. Syrah is cultivated in around 100 hectares in several regions and states, such as Minas Gerais, São Paulo, Rio de Janeiro, and Espírito Santo, in the Brazilian South-East, in a sub-tropical climate. Subsequently, it was introduced in Goiás, Mato Grosso, and Federal District, in the Center-West, also in a sub-tropical climate, and since 2012 it is being cultivated in the Chapada Diamantina, in Bahia state, in a tropical of altitude climate. In all of these regions, the main factor is the high altitudes where vineyards are planted, ranging from 700 to 1.200 m above sea level. The principal conditions during the harvest season, between June and August, are warm days (between 18-25°C), cold nights (ranging between 2-4°C to 8-10°C), with high thermal amplitude, very low or null pluviosity, and blue sky, promoting a high enological potential of the grapes.

The main characteristic of this new viticulture in Brazil is the technique of double pruning, where the normal cycle starts with pruning in August (named formation pruning, with just one bud per spur), as it is made in all viticultural countries of the South Hemisphere. But in October-November, the grape bunches are eliminated, and the branches are ripened in December. Then, the second pruning (named the production pruning) is carried out in January, hydrogen cyanamide is applied to promote and homogenize the budburst, and it starts the second cycle [13, 43-46]. The harvest period is between June and August, depending on the grape variety (early, intermediate, or late), being considered ideal to produce high-quality Syrah winter wines [43-47]. In contrast, other varieties are being used successfully, such as Cabernet Sauvignon, Cabernet Franc, Malbec, Petit Verdot, for reds, and Sauvignon Blanc and Viognier, for whites [13].

Phenolic compounds were determined in Syrah wines from different winegrowing regions in Brazil, in traditional wines (Santa Catarina), tropical wines (São Francisco Valley), and winter wines (Minas Gerais, São Paulo, Chapada Diamantina), presenting significative variations (Table 3). It is interesting to note that Syrah wines from Minas Gerais and São Paulo presented the highest concentrations of total phenols, as compared to Syrah wines from Santa Catarina and São Francisco Valley. Also, winter wines from São Paulo showed higher total anthocyanins than wines from Santa Catarina, Minas Gerais, and São Francisco Valley. Wines presented different concentrations of other phenolics (Table 3).

A comparison was carried out between Syrah wines from a tropical semi-arid climate (tropical wines, produced at 350 m a.s.l.) and tropical of altitude climate (winter wines, produced at 1.100 m a.s.l.), both regions located in the Brazilian North-East (São Francisco Valley and Chapada Diamantina) [48]. Results showed that the altitude and its peculiarities (day/night thermal amplitude, precipitation, and solar radiation) had a significant effect on the chemical composition of the wines, primarily concerning the phenolic compounds and sensorial profiles (Table 4).

Table 3. Phenolic composition of Syrah red wines from different regions of Brazil

Compound	Concentration	Region	Reference
Total Phenols (mg.L^{-1})			
	2376-3009	Minas Gerais	[10]
	1176-2434	Santa Catarina	[8]
	1068 - 2313	São Francisco Valley	[48]
	2410-3300	São Paulo	[12]
Total monomeric anthocyanins (mg.L^{-1})			
	95 – 126	Minas Gerais	[49]
	27 – 134	Santa Catarina	[8]
	38 - 244	São Francisco Valley	[50]
	415-550	São Paulo	[12]
Catechin (mg.L^{-1})			
	94.6	São Francisco Valley	[51]
	3.5 - 26.1	São Francisco Valley	[9]
	2.9-4.3	Minas Gerais	[10]
Condensed tannins (mg.L^{-1})			
	443-1037	São Francisco Valley	[48]
	419 - 485	Chapada Diamantina	[48]
Trans-resveratrol (mg.L^{-1})			
	0.6 - 6.6	Santa Catarina	[8]
	7.41	São Francisco Valley	[9]
	10.5	São Francisco Valley	[51]
	2.2 - 3.5	Chapada Diamantina	[48]
	0.4 - 0.5	Minas Gerais	[10]

Syrah winter wines presented higher levels of total phenols, flavonoids, total and colored anthocyanins, color intensity, monomeric and oligomeric tannins, trans-resveratrol, anthocyanins 3-*O*-glucoside (peonidin, petunidin, and malvidin), and procyanidins B3, B1 3-O-gallate, and Trimer 2. Syrah tropical wines presented higher levels of quercetin, isorhamnetin, total flavonols, polymeric tannins, and anthocyanins 3-O-acetylglucoside (petunidin and malvidin) [48].

Table 4. Phenolic composition of Syrah wines produced in the São Francisco Valley (tropical wines) and in the Chapada Diamantina-Bahia state (winter wines) [48]

Region	1100 m altitude		350 m altitude		Sig.
Harvest	2014	2015	2014	2015	
Phenolic compounds					
Total phenols (mg L^{-1})	2142.9^a	1676.9^b	1456.9^c	1401.8^c	***
Non-flavonoids (mg L^{-1})	162.4^c	194.7^a	185.0^b	155.9^d	**
Flavonoids (mg L^{-1})	1981.1^a	1486.0^b	1272.4^c	1246.5^d	***
Color, anthocyanins and other pigments					
Total anthocyanins (mg L^{-1})	372.7^a	340.4^b	237.5^d	314.3^c	***
Colored anthocyanins (mg L^{-1})	74.3^a	50.5^b	35.1^d	38.4^c	***
Ionization index (%)	19.9^a	14.8^b	14.5^b	11.7^c	***
Total pigments (u.a)	13.1^a	11.6^b	9.2^d	10.6^c	**
Polymerized pigment (u.a)	4.2^a	2.9^b	1.8^c	1.9^c	**
Polymeric pigments index (%)	32.1^a	25.0^b	19.6^c	17.9^d	**
Condensed tannins (mg L^{-1})					
Monomeric	23.6^a	23.9^a	16.3^b	12.6^c	***
Oligomeric	159.1^a	146.6^b	142.5^{bc}	139.3^c	**
Polymeric	302.8^c	249.0^d	665.4^a	400.0^b	***
Total tannins	485.5^c	419.5^d	824.3^a	551.9^b	***
Tannin power (NTU mL^{-1})	191.0^c	153.2^d	212.1^a	203.2^b	***

Means followed by the same letter in the lines did not differ by Tukey test at 5% (p ≤ 0.05). Standard deviation of triplicate analysis. Total phenols, non-flavonoids and flavonoids expressed as mg L^{-1} of gallic acid; anthocyanins expressed as mg L^{-1} of malvidin; u.a (absorbance unit); n.s. (not significant); * (significant differences at a 95% confidence level); ** (significant differences at a 99.9% confidence level); *** (significant differences at a 99.99% confidence level); 1100m (Bahia); 350m (Pernambuco).

Sensory Profile of Syrah Wines

Wine grapes grown in different climatic and geographical areas led to a diverse expression of varietal characteristics [52]. The Syrah red wine has a distinctive aromatic profile, which is usually characterized by the aroma of red fruits and berry-like, also often described with spices or black pepper attributes.

Scientific studies have shown that the composition of Syrah wines produced in different regions of Brazil, presented particular typicalities, but in general, they are similar in terms of quality, as compared to those wines produced in other traditional regions from temperate climates, such as France, Italy, Australia, Spain, and other countries. Young winter wines, produced in the Southwest region of Brazil, present a high intensity of color, are rich in ester and monoterpene, as well as alcoholic volatile compounds responsible for ethereal, fruity, flowery, fresh, and sweet aromas. The winter wines (aged/guard) were characterized with higher contents of furfural, geranyl ethyl ether, isoamyl decanoate, α-murolene, and α-calacorene, contributing to sweet, fruity, and woody aromas [53]. Some aged/guard wines from South-East and Chapada Diamantina present high color intensity, purple/violet, with notes of vanilla, flowers such as violet, fresh, and very complex. The taste is fresh, with good acidity, full-bodied, concentrated, with freshness and persistence.

In the São Francisco Valley, the sensory profile of red Syrah tropical wines was characterized by the presence of fruity aroma, with notes of floral and empyreumatic (spices, toasted) [11, 54]. The compounds hexanol, ethyl dodecanoate, ethyl decanoate, 1-propanol, 2-phenylethanol, 2-phenyl-ethyl acetate, 3,4 dimethyl 2-hexanol were also identified (Table 5) [55, 56].

A study in the same region with traditional sparkling wines elaborated with Syrah demonstrated that the volatile profile is related to compounds such as benzaldehyde, butanoic acid, isoamyl acetate, (Z)-3-hexen-1-ol, cis-3-hexen-1-ol acetate, and hexyl acetate, whose flavors are associated to sweet aromatic notes, butter, cheese, fruity, vegetable, oily, cherry, and pear [57].

According to the taste analysis, Syrah tropical red wines produced in the São Francisco Valley are generally characterized as acidic, astringent (tannic), with high alcoholic potential, and persistent; however, it depends on the harvest date throughout the year [41].

**Table 5. Main aromatic compounds identified
in tropical red wines of Syrah variety [55-57]**

Chemical compounds	Aromatic descriptors
Ethyl decanoate	Fruity (grape)
2-Phenyl ethyl acetate	Flower, rose
1-Propanol	Flower, fruity
1-Hexanol	Vegetable, grass
Rotundone	Peppery
2-phenylethanol	Rose

Influence of Winemaking Technology on Wine Characteristics

An increasing number of new winemaking practices are currently being used by winemakers worldwide. Several studies have reported that different winemaking technologies can alter strongly the composition of the wine and its sensory profile, depending on the variety [58-61]. For red wines, the traditional winemaking process can extract only a fraction of the large amounts of different phenolic compounds, located in the grape skins, due to the resistance to mass transfer of cell walls and cytoplasmic membranes [59]. Different winemaking techniques have been developed, such as an increase of the fermentation temperature, thermovinification, time of maceration, freezing of must or grapes, types of yeasts, and the use of enzymes in maceration to improve the extraction of phenolic compounds and, consequently, the quality of red wines [60, 61]. Other studies with different alternative pretreatments, such as ultrasound, pulsed electric fields, and high voltage electrical discharges, have been also evaluated as alternatives to increase the extraction of phenolic compounds from grape berries, skins, and seeds [62-64].

Studies with Syrah wines have been shown, such as the extraction factor and the behavior of phenolic compounds during maceration at different temperatures and times. Gómez-Míguez et al. [65] reported higher extraction rates of anthocyanins and other phenolic compounds in wines elaborated with cold maceration. Cejudo-Bastante et al. [66] verified that phenolic compounds showed significantly higher content in cold macerated wines when longer contact time was used. The use of cold maceration tests, followed by no plunge, and only no plunge, showed that extractions of total phenolic total, tannins, and colors were more stable over time [67].

In must freezing, grapes are kept at negative temperatures and extraction of phenols takes place in the absence of ethanol. This technique can be used as an alternative method to increase red wine pigments, tannins, and aroma, because the oxidation of anthocyanin and aroma compounds is delayed or prevented [68, 69].

Ghanem et al. [70], evaluating macerations at different temperatures (10, 60, 70, and 80°C) for 48 h, reported that the pre-fermentation heat treatment of Syrah grapes is more efficient for the extraction of polyphenols than the cold maceration. An analysis of must samples revealed a systematic increase in the concentration of tannins with temperature and over time. Temperature favored anthocyanin extraction, but the degradation of these compounds was observed at high temperatures when the maceration was extended beyond eight hours [70].

A study with the addition of enzymes in the maceration process concluded that it favors a greater intensity of color intensity and specifically red color, decreased luminosity, and increased content of proanthocyanidins in Syrah wines [71]. Other authors reported that the addition of enzyme resulted in a significantly higher concentration of tannin, which increased wine astringency. Conversely, mannoprotein addition reduced tannin concentration and astringency. The addition of oenotannin did not influence wine composition or sensory properties [34].

Studies developed by Pereira et al. [15], evaluating and comparing manual and mechanical destemming of grapes influenced the chemical and sensorial composition of Syrah tropical wines in the São Francisco Valley. As explained before, the vine cycle is too fast in that region, and it was observed that the stem/rachis of some varieties remains green at harvest period. The authors obtained higher concentrations of total anthocyanins in Syrah wines elaborated from manual destemming, in which total tannins were higher in wines from mechanical destemming. The sensorial analysis showed that the wines from manual obtained the higher scores of color intensity, fruity and floral aromas, taste intensity, while wines from mechanical treatment were characterized as astringent and bitter [15]. It was suggested to wineries located in the tropical semi-arid region of Northeastern Brazil the use of belt sorting after destemming, or even adopt the manual destemming for high-quality red wines. Another alternative is the use of mechanical harvest since it was observed in the region a smaller quantity of stems during winemaking with a machine, but further studies need to be carried out to confirm this hypothesis.

The final aroma of wine is originated from both the grape and the fermentation process, but there are some compounds in the grapes that are responsible for the final varietal character of the wines. Ruiz-Rodríguez et al. [72] evaluating the application of ultrasound during the fermentation process, concluded that this treatment favored the extraction of volatile compounds and increased astringency in Syrah wines. Another study, also with the use of ultrasound, demonstrated that this method increased the extraction of total phenolic compounds and anthocyanins [73].

FINAL REMARKS

The socioeconomic importance of the Syrah variety to the world vitiviniculture is notable in both traditional as well as in new regions. It has a great diversity of the agronomical grapevine responses according to the terroir, but also different styles of wines produced, mainly used for red wines (young or aged/guard), but also for sparklings in Brazil.

In Brazil, the only country where it is possible to have three kinds of viticulture and wines in the world, the facility of Syrah grapevine adaptation to different climates, soils, and management has allowed the development of new viticultural zones. It has been used in the first and the most traditional winegrowing zone of the country, in temperate climates in Southern, with very interesting red wines, young and aged/guard. In the second viticulture, located in Northeastern, tropical wines have been produced with Syrah, for young and guard wines, but also for very interesting sparkling wines, among whites and rosés, in blends with white and red varieties. Finally, a third and the newest winegrowing zone in Brazil, in Southeastern, Centerwestern and Northeastern regions, all of them around 1.000 m of altitude, where winter wines are being produced with high success, and Syrah is used to make young and aged red wines, with other traditional varieties used. The most important is that the Syrah grapevine is a global grape variety, which adapted successfully in all winegrowing zones of Brazil, where red wines and sparkling wines can be found, and all products present quality and typicality.

REFERENCES

[1] Mullins, M.G., A. Bouquet, & L.E. Williams. (1992). *Biology of the grapevine.* Cambridge University Press, Cambridge, United Kingdom.

[2] Organisation Internationale de la Vigne et du Vin, Focus OIV. (2019). *State of the world vitivinicultural – Sector in 2019* report.

[3] *Anivin de France, Vin de France & cepages, Encyclopédie dés Cepages de France*, OIV, 2019. [*Anivin de France, Vin de France & grape varieties, Encyclopedia of grape varieties of France*]

[4] Yoo, Y.J., Prenzler, P.D., Saliba, A.J., & Ryan, D. (2011). Assessment of Some Australian Red Wines for Price, Phenolic Content, Antioxidant Activity, and Vintage in Relation to Functional Food Prospects. *Journal of Food Science, 76*(9), C1355-C1364.

[5] Bimpilas, A., Tsimogiannis, D., Balta-Brouma, K., Lymperopoulou, T., & Oreopoulou, V. (2015). Evolution of phenolic compounds and metal content of wine during alcoholic fermentation and storage. *Food Chemistry, 178*, 164–171.

[6] Barrio-Galán, R., Medel-Marabolí, M., & Peña-Neira, A. (2015). Effect of different aging techniques on the polysaccharide and

phenolic composition and sensory characteristics of Syrah red wines fermented using different yeast strains. *Food Chemistry, 179*, 116–126.

[7] Lingua, M.S., Fabani, M.P., Wunderlin, D.A., & Baroni, M.V. (2016). In vivo antioxidant activity of grape, pomace and wine from three red varieties grown in Argentina: Its relationship to phenolic profile. *Journal of Functional Foods, 20*, 332–345.

[8] Sartor, S., Malinovski, L.I., Caliari, V., Silva, A.L., & Bordignon-Luiz, M.T. (2017). Particularities of Syrah wines from different growing regions of Southern Brazil: grapevine phenology and bioactive compounds. *Journal of Food Science and Technology, 54*(6), 1414–1424.

[9] Belmiro, T.M.C., Pereira, C.F., & Paim, A.P.S. (2017). Red wines from South America: Content of phenolic compounds and chemometric distinction by origin. *Microchemical Journal, 133*, 114-120.

[10] Vilas-Boas, A.C., Nassur, R.C.M.R., Henrique, P.C., Pereira, G.E., & Lima, L.C.O. (2018). Bioactive compounds in wines produced in a new area for vitiviniculture in Brazil. *Bioscience Journal, 35*(5), 1356-1368.

[11] Oliveira, J.B., Egipto, R., Laureano, O., Castro, R. de, Pereira, G.E., & Ricardo-da-Silva, J.M. (2019a). Chemical composition and sensory profile of Syrah wines from semiarid tropical Brazil – Rootstock and harvest season effects. *LWT - Food Science and Technology, 114*, 1-9.

[12] Mota, R.V., Peregrino, I., Rivera, S.P.T., Hassimotto, N.M.A., de Souza, A.L., & de Souza, C.R. (2019). Characterization of Brazilian Syrah winter wines at bottling and after ageing. *Scientia Agricola, 78*, 1-10.

[13] Pereira, G.E. (2020). The three different winegrowing zones in Brazil according to climate conditions and vine managements. In: Jordão and Botelho, Vitis: *Biology and Species*. Available at: https://novapublishers.com/shop/vitis-biology-and-species/.

[14] Tonieto, J., & Pereira, G.E. (2012). *A concept for the viticulture of tropical wines*. In: *Proceedings of the IXth International Terroir Congress*, 2012, p. 34-37.

[15] Pereira, G. E., Guerra, C. C., Amorim, F. F., Nascimento, A. M. S., Souza, J. F., Lima, L. L. A., Lima, M.S., Padilha, C.V.S., Protas, J.F.S, Zanus, M.C., & Tonietto, J. (2018). Semi-Arid tropical wines of Brazil. Discovering the wine potential of this new geographic frontier of wine. *Territoires du Vin, 9*, 1–13.

[16] Robinson J. (2006). *The Oxford Companion to Wine*, Third Edition, p. 676-677, Oxford University Press, 676-677.

[17] Vouillamoz, J.F., & Grando, M.S. (2006). Genealogy of wine grape cultivars: Pinot is related to Syrah. *Heredity, 97*, 102-110.

[18] Organisation Internationale de la Vigne et du Vin, Focus OIV. 2017. *Distribution variétale du vignoble dans le monde*. [*Varietal distribution of the vineyard in the world.*]

[19] Robinson, J. 1986. *Vines, Grapes & Wine*. Octopus Publishing, 90.

[20] Visan, L., Tamba-berehoiu, R., Popa, C.N., Danaila-Guidea, S.N., & Dobrinoiu, R.V. (2019). Syrah – Grapevine and wine - A critical review. *Scientific Papers Series Management, Economic Engineering in Agriculture and Rural Development, 19*, 609-615.

[21] Günata, Z., Wirth, J., Guo, W., & Baumes, R. (2002). C13-Norisoprenoid aglycon composition of leaves and grape berries from Muscat of Alexandria and Shiraz cultivars. In P. Winterhalter & R. Rouseff (Orgs.), *Carotenoid-derived aroma compounds* (Vol. 802, p. 255–261). ACS Publications.

[22] Garrido, J., & Borges, F. (2013). Wine and grape polyphenols - A chemical perspective. *Food Research International, 54*, 1844-1858.

[23] Van Leeuwen, C., Friant, P., Choné, X., Tregoat, O., Koundouras, S., &Dubourdieu, D. (2004). Influence of climate, soil, and cultivar on terroir. *American Journal of Enology and Viticulture, 55*(3), 207-217.

[24] Van Leeuwen A. & Seguin G. (2006). The concept of terroir in viticulture. *Journal of Wine Research, 16*, 01-10.

[25] Reynier, A. (2007). *Manuel de viticulture,* 10 ed. Paris: Editions Tec et Doc Lavoisier. [*Viticulture manual*]

[26] Usseglio-Tomasset, L. 1995. *Chimie oenologique.* 2. ed. Paris: Lavoisier Techniqe et Documentation, 387. [*Oenological chemistry*]

[27] Peynaud, E. (1997). *Connaissance et travail du vin.* 2. ed. Paris: Dunod, 341. [*Knowledge and work of wine*]

[28] Ribéreau-Gayon, P., Glories, Y., Maujean, A., & Dubourdieu, D. (2000). *Handbook of enology. The chemistry of wine, stabilization and treatments.* 2 ed. Wiley.

[29] Casassa, L.F., & Harbertson, J.F. (2014). Extraction, evolution, and sensory impact of phenolic compounds during red wine maceration. *Annual Review of Food Science and Technology, 5,* 83-109.

[30] Rivero, F.J., Gordillo, B., Jara-Palacios, M.J., González-Miret, L.M., & Heredia, F.J. (2017). Effect of addition of overripe seeds from white grape by-products during red wine fermentation on wine colour and phenolic composition. *LWT - Food Science and Technology, 84,* 544-550.

[31] Ristic, R., Cozzolino, D., Jeffery, D.W., Gambetta, J.M., & Bastian, S.E.P. (2016). Prediction of phenolic composition of Shiraz wines using attenuated total reflectance mid-infrared (ATR-MIR) spectroscopy. *American Journal of Enology and Viticulture,* doi: 10.5344/ajev.2016.16030.

[32] Šuklje, K., Zhang, X., Antalick, G., Clark, A.C., Deloire, A., &Schmidtke, L.M. (2016). Berry shriveling significantly alters Shiraz (*Vitis vinifera* L.) grape and wine chemical composition. *Journal of Agricultural and Food Chemistry, 64*(4), 870–880.

[33] Gil-Muñoz, R., Bautista-Ortín, A.B., Ruiz-García, Y., Fernández-Fernández, J.I., & Gómez-Plaza, E. (2017). Improving phenolic and chromatic characteristics of Monastrell, Merlot and Syrah wines by using methyl jasmonate and benzothiadiazole. *OENO One, 51,* 17-27.

[34] Li, S., Bindon, K.A., Bastian, S.E.P., Jiranek, V., & Wilkinson, K.L. (2017). Use of winemaking supplements to modify the composition

and sensory properties of Shiraz wine. *Journal of Agricultural and Food Chemistry, 65*(7), 1353–1364.

[35] Favre, G., Piccardo, D., Sergio, G.-A., Pérez-Navarro, J., García-Romero, E., Mena-Morales, A., & González-Neves, G. (2020). Stilbenes in grapes and wines of Tannat, Marselan and Syrah from Uruguay. *OENO One, 54*(1), 27-36.

[36] Vitalini, S., Gardana, C., Simonetti, P., Fico, G., & Iriti, M. (2012) Melatonin, melatonin isomers and stilbenes in Italian traditional grape products and their antiradical capacity. *Journal of Pineal Research, 54*, 322–333.

[37] Harbertson, J.F., Hodgins, R.E., Thurston, L.N., Schaffer, L.J., Reid, M.S., Landon, J.L., Ross, C.F., & Adams, D.O. (2008). Variability of tannin concentration in red wines. *American Journal of Enology and Viticulture, 59*(2), 210-214.

[38] Cosme, F, Ricardo-da-Silva, J.M., & Laureano, O. (2009). Tannin profiles of Vitis vinifera L. cv. red grapes growing in Lisbon and from their monovarietal wines. *Food Chemistry, 112*, 197–204.

[39] Mello, L.M.R. de. (2019). *Vitivinicultura brasileira*: panorama 2018. Embrapa Uva e Vinho. *Comunicado técnico, 210*: 1-10. Available at: http://ainfo.cnptia.embrapa.br/digital/bitstream/item/203100/1/Comunicado-Tecnico-210.pdf.

[40] Pereira, G. E., Tonietto, J., Zanus, M. C., Santos, H. P., Protas, J. F. S., Mello, L. M. (2020). *Vinhos no Brasil: contrastes na geografia e no manejo das videiras nas três macrorregiões climáticas do país.* Embrapa: Documentos. https://www.embrapa.br/en/uva-e-vinho/publicacoes. [*Wines in Brazil: contrasts in geography and grapevine management in the country's three climatic macro-regions.*]

[41] Pereira, G. E., Padilha, C., Biasoto, A. C. T., Canuto, K. M., Nascimento, A. M. S., & De Souza, J. F. (2016). Le poids des consommateurs sur l'evolution des vins: L'exemple de la Vallee du Sao Francisco, Brésil. In J. Perard & M. Perrot (Orgs.), *Le poids des consommateurs sur l'evolution des vins: L'exemple de la Vallee du Sao Francisco, Brésil* (1º ed, p. 301–310). Centre Georges Chevrier. [The influence of consumers on the evolution of wines: The example

of the Sao Francisco Valley, Brazil. In J. Perard & M. Perrot (Orgs.), *The influence of consumers on the evolution of wines: The example of the Sao Francisco Valley, Brazil*]

[42] Oliveira, J.B., Egipto, R., Laureano, O., Castro, R. de, Pereira, G.E., & Ricardo-da-Silva, J.M. (2019b). Chemical characteristics of grapes cv. Syrah (*Vitis vinifera* L.) grown in the tropical semiarid region of Brazil (Pernambuco state): influenceof rootstock and harvest season. *Journal of the Science of Food and Agriculture, 99*, 5050–506.

[43] Favero, A.C., Amorim, D.A., Mota, R.V., Soares, A.M., Souza, C.R., & Regina, M.A. (2011). Double-pruning of 'Syrah' grapevines: a management strategy to harvest wine grapes during the winter in the Brazilian to harvest wine grapes during the winter in the Brazilian Southeast. *Vitis, 50*(4), 151-158.

[44] Regina, M.A., Mota, R.V., Souza, C.R., & Favero, A.C. (2011). Viticulture for the wines in Brazilian Southeast. *Acta Horticulturae, 910*, 113-120.

[45] Dias, F.A.N., Mota, R.V., Souza, C.R., Pimentel, R.M.A., Souza, L.C., Souza, A.L., & Regina, M.A. (2017). Rootstock on vine performance and wine quality of 'Syrah' underdoublepruning. *Scientia Agricola, 74*(2), 134-141.

[46] Almeira Junior, O. de, Souza, C.R. de., Dias, F.A.N., Fernandes, F.de P., Torregrosa, L., Fernandes-Brum, C.N., Chalfun Junior, A., Mota, R.V. da, Peregrino, I., & Regina, M.de A. (2019). Effect of pruning strategy on 'Syrah' bud necrosis and fruitfulness in Brazilian subtropical Southeast. *Vitis, 58*(3), 87-94.

[47] Wurz, D. A., de Bem, B. P., Allebrandt, R., Bonin, B., Dalmolin, L.G., Canossa, A.T., Rufato, L., & Kretzschmar, A.A. (2017). New wine-growing regions of Brazil and their importance in the evolution of Brazilian wine. *BIO Web of Conferences 9*, 1- 4.

[48] Oliveira, J.B., Egipto, R., Laureano, O., Castro, R. de, Pereira, G.E., & Ricardo-da-Silva, J.M. (2019c). Chemical and sensorial characterization of tropical Syrah wines produced at different

altitudes in northeast of the Brazil. *South African Journal of Enology and Viticulture, 40*(2), 157-171.

[49] Mota, R. V., Souza, C. R., Pereira, G. E., Dal'osto, M. C., Correa, L. C., Dias, F. A. N., Pimentel, R. M. A., Tavares, M. C., & Regina, M. A. (2012). Characterization of grapes and wines of Syrah growing during the autumn-winter season in different viticultural zones of the Brazilian Southeast. *Annals of the IX Congres des terroirs vitivinicoles, 9*, 45–48.

[50] Souza, J.F., Nascimento, A.M.S., Linhares, M.S.S., Dutra, M.C.P., Lima, M.S., & Pereira, G.E. (2018). Evolution of phenolic compound profiles and antioxidant activity of Syrah red and sparkling Moscatel wines stored in bottles of different colors. *Beverages, 89*(4), 1-13.

[51] Mucaca, C.A.L., Filho, J.H.T., Nascimento, E., & Arruda, L.L.A.L. (2017). Phenolic composition, chromatic parameters and antioxidant activity "in vitro" in Tropical Brazilian red wines. *Journal of Food and Nutrition Research, 10*, 754-762.

[52] Pérez-Magariño, S., Ortega-Heras, M., Bueno-Herrera, M., Martínez-Lapuente, L., Guadalupe, Z., &Ayestarán, B. (2014). Grape variety, aging on lees and aging in bottle after disgorging influence on volatile composition and foamability of sparkling wines. *LWT - Food Science and Technology, 61*, 47–55.

[52] Mota, R.V., Peregrino, I., Rivera, S.P.T., Hassimoto, N.M.A., Souza, A.L., & Souza, C.R. (2020). Characterization of Brazilian Syrah winter wines at bottling and after ageing. *Scientia Agricola, 78*(3), 1-10.

[54] Oliveira, J. B., Silva, G. G., Araújo, A. J. B., Lima, L. L. A., Ono, E. O., Castro, R., Cruz, A., Santos, J., & Pereira, G. E. (2012). *Influence of the vintage, clone and rootstock on the chemical characteristics of Syrah tropical wines from Brazil.* 17–20.

[55] Araújo, A. J. B., Vanderlinde, R., Oliveira, J. B., Oliveira, G. G., Biasoto, A. C. T., & Pereira, G. E. (2012). *Aromatic stability of Syrah and Petit Verdot tropical wines from Brazil. 9*, 6–9.

[56] Barbará, J. A., Biasoto, A. C. T., Nicolli, K. P., Queiroz, L. B., Silva, E. A. S., Correa, L. C., Welke, J. E., & Zini, C. A. (2015). Evaluating the typicality of the aroma of Syrah tropical wines from the Sub-middle São Francisco Valley employing gas chromategraphy coupled to olfactometry detection (GC-O). *Annals of the XV Congresso latino-americano de viticultura e enologia, 15.*

[57] Nascimento, A.M.S., Souza, J.F., Lima, M.S., & Pereira, G.E. (2018) Volatile profiles of sparkling wines produced by the traditional method from a Semi-Arid region. *Beverages, 4*(103), 1-12.

[58] Smith, P.A, McRae, J.M., & Bindon, K.A. (2015). Impact of winemaking practices on the concentration and composition of tannins in red wine. *Australian Journal of Grape and Wine Research, 21*, 601–614.

[59] Donsìa F., Ferraria G., Fruilob M., & Pataroa G. (2011). Pulsed Electric Fields – assisted vinification. *Procedia Food Science, 1*, 780-785.

[60] Sacchi, K.L., Bisson, L.F., & Adams, D.O. (2005). A review of the effect of winemaking techniques on phenolic extraction in red wines. *American Journal of Enology and Viticulture, 56*, 197–206.

[61] Moreno-Arribas, M.V., & Polo, M.C. (2005). Winemaking biochemistry and microbiology: Current knowledge and future trends. *Critical Reviews of Food Science and Nutrition, 45*, 265–286.

[62] Delsart, C., Ghidossi, R., Poupot, C., Cholet, C., Grimi, N., Vorobiev, E., Milisic, V., & Peuchot, M. (2012). Enhanced extraction of valuable compounds from merlot grapes by pulsed electric field. *American Journal of Enology and Viticulture,* 11088.

[63] Boussetta, N., Vorobiev, E., Deloison, V., Pochez, F., Cordin-Falcimaigne, A., & Lanoiselle, J.L. (2011). Valorisation of grape pomace by the extraction of phenolic antioxidants: application of High-Voltage Electrical discharges. *Food Chemistry, 128*, 364–370.

[64] López-Giral, N., González-Arenzana, L., González-Ferrero C., López R., Santamaría P., López-Alfaro, I., & Garde-Cerdán, T. (2015) Pulsed electric field treatment to improve the phenolic compound extraction from Graciano, Tempranillo and Grenache

grape varieties during two vintages. *Innovative Food Science and Emerging Technologies.* Available at: http://dx.doi.org/10.1016/j.ifset.2015.01.003.

[65] Gómez-Míguez, M., González-Miret, M.L., & Heredia, F.J. (2007). Evolution of colour and anthocyanin composition of Syrah wines elaborated with pre-fermentative cold maceration. *Journal of Food Engineering, 79*, 271-278.

[66] Cejudo-Bastante, M.J., Gordillo, B., Hernanz, D., Escudero-Gilete, M.L., González-Miret, M.L., & Heredia, F.J. (2014). Effect of the time of cold maceration on the evolution of phenolic compounds and colour of Syrah wines elaborated in warm climate. *International Journal of Food Science and Technology*, 1-7.

[67] Chittenden, R., & King, P. (2020). No plunging and cold maceration followed by no plunging as alternative winemaking techniques: tannin extraction and pigment composition of Syrah and Pinot Noir wines. *South African Journal for Enology and Viticulture, 41*(1), 90-98.

[68] Gil-Muñoz, R., Moreno-Perez, A., Vila-Lopez, R., Fernandez-Fernandez, J.I., Martinez-Cutillas, A. & Gomez-Plaza, E. (2009). Influence of lowtemperatureprefermentative techniques on chromatic and phenoliccharacteristics of Syrah and Cabernet Sauvignon wines. *European Food Research and Technology, 228*, 777-788.

[69] Gordillo, B., Cejudo-Bastante, M.J., Rodríguez-Pulido, F.J., Jara-Palacios, M.J., Ramírez-Pérez, P., González-Miret, M.L., & Heredia, F.J. (2014). Impact of adding white pomace to red grapes on the phenolic composition and color stability of Syrah wines from a warm climate. *Journal of Agricultural and Food Chemistry, 62*(12), 2663–2671.

[70] Ghanem, C., Taillandier, P., Rizk, Z., Nehme, N., Souchard, J.P., & El Rayess, Y. (2019). Evolution of polyphenols during Syrah grapes maceration: Time versus temperature effect. *Molecules, 24*(15), 1-13.

[71] González-Neves, G., Gil, G., Favre, G., Baldi, C., Hernández, N., & Traverso, S. (2013). Influence of winemaking procedure and grape

variety on the colour and composition of young red wines. *South African Journal for Enology and Viticulture, 34*(1), 138-146.

[72] Ruiz-Rodríguez, A., Carrera, C., Lovillo, M.P., & Barroso, C.G. (2019). Ultrasonic treatments during the alcoholic fermentation of red wines: effects on Syrah wines. *Vitis, 58*, 83–88.

[73] Mazza, K.E., Santiago, M.C., do Nascimento, L.S., Godoy, R. L., Souza, E.F., Brígida, A.I.S., & Tonon, R.V. (2018). Syrah grape skin valorisation using ultrasound-assisted extraction: Phenolic compounds recovery, antioxidant capacity and phenolic profile. *International Journal of Food Science & Technology, 54*, 641-650.

About the Editors

Maurício Bonatto Machado de Castilhos, PhD in Food Engineering and Science at the São Paulo State University (UNESP) (2016), São José do Rio Preto, Brazil. Currently, he holds the position of Professor and Researcher at Minas Gerais State University (UEMG), Frutal, Brazil. He researches the physicochemical and sensory profile of fermented beverages, primarily wines, submitted to alternative winemaking processes. Also, he researches on identification and quantitation of phenolic and volatile compounds by High-Performance Liquid Chromatography coupled with Mass Spectrometry (HPLC-MS^n) and Gas Chromatography (GC-MS), respectively. He presents basic knowledge in wine metabolomics with the application of the Nuclear Magnetic Resonance technique (NMR).

About the Editors

Vanildo Luiz Del Bianchi, Assistant Professor Doctor at the Universidade Estadual Paulista - UNESP, São José do Rio Preto. PhD in Agronomy (Energy in Agriculture) at São Paulo State University (1998). He has experience in the Food Science and Technology area, with emphasis on Biochemical Engineering, working primarily on the following topics: Treatment (Aerobic and Anaerobic) of Agroindustrial Wastewater, Use of Agroindustrial Waste, Bioprocesses, Fermentation in Solid State, Production of Enzymes, Bioinsecticides, Fermented Beverages, and Biosurfactants.

Vitor Manfroi, Enologist at the Federal Agrotechnical School of Bento Gonçalves (1982), PhD in Agroindustrial Science and Technology from the Federal University of Pelotas (2007). He is an Associate Professor at the Federal University of Rio Grande do Sul, and he was Director of the Institute of Food Science and Technology (ICTA/UFRGS) from 2010 to 2018. He has extensive experience in Beverage Technology, working primarily on the following topics: enology (wines and sparkling wines), juice technology, distillate technology, and beer technology, with a current focus on process technology and good manufacturing practices.

INDEX

A

acetic acid, 7, 56, 68, 97, 146, 148
acid, 6, 12, 26, 38, 39, 54, 55, 57, 58, 59, 60, 61, 62, 63, 66, 67, 69, 71, 73, 97, 98, 100, 119, 122, 142, 144, 148, 152, 158, 172, 180, 212, 213
acidity, 3, 6, 14, 54, 57, 59, 60, 63, 68, 69, 71, 85, 117, 119, 120, 122, 142, 153, 176, 203, 213
active compound, 59, 95, 96, 97, 110
aging, 2, 4, 12, 22, 23, 37, 43, 44, 46, 80, 81, 89, 90, 91, 92, 94, 101, 115, 118, 139, 164, 165, 166, 168, 169, 170, 171, 173, 182, 184, 185, 188, 193, 195, 201, 204, 217, 223
aging process, 92, 182, 204
air temperature, 120
alcohols, 11, 12, 13, 15, 60, 62, 69, 110, 134, 135, 145, 146, 180, 181, 182
anthocyanin, 15, 31, 32, 34, 37, 49, 50, 89, 98, 100, 120, 124, 128, 141, 157, 158, 166, 167, 168, 169, 170, 171, 187, 189, 190, 215, 225

anthocyanins, vii, 6, 10, 13, 14, 22, 30, 31, 32, 33, 34, 36, 37, 38, 39, 46, 50, 81, 83, 88, 89, 90, 91, 98, 99, 101, 107, 108, 112, 119, 120, 122, 123, 127, 128, 129, 132, 139, 141, 142, 143, 144, 152, 153, 158, 164, 165, 166, 167, 168, 169, 170, 176, 177, 189, 190, 204, 207, 210, 211, 212, 214, 215, 216
antioxidant, viii, 2, 4, 11, 13, 14, 15, 19, 49, 51, 83, 108, 114, 157, 166, 188, 218, 223, 226
antioxidant capacity, viii, 2, 14, 15, 51, 114, 157, 166, 188, 226
aroma, vi, vii, viii, 2, 3, 11, 12, 13, 17, 45, 55, 59, 60, 61, 62, 70, 72, 73, 74, 75, 76, 77, 80, 81, 82, 89, 91, 92, 94, 95, 96, 97, 104, 110, 111, 115, 116, 118, 132, 133, 134, 138, 141, 146, 149, 150, 151, 152, 153, 159, 160, 161, 162, 164, 165, 174, 175, 178, 179, 180, 181, 182, 184, 185, 192, 193, 194, 195, 196, 213, 215, 216, 219, 224
aroma composition, vi, 115, 132, 153, 194

aroma compounds, 55, 116, 133, 134, 146, 150, 153, 174, 179, 185, 192, 193, 195, 215, 219
aroma precursors, 73, 164, 175, 178, 185
aromatic alcohols, 134, 135
aromatic compounds, 44, 60, 76, 103, 155, 168, 175, 177, 214
astringency, 14, 15, 22, 28, 29, 30, 44, 45, 46, 49, 51, 72, 80, 81, 86, 88, 97, 98, 100, 101, 102, 112, 113, 141, 143, 165, 172, 173, 185, 191, 204, 215, 216
astringent, 25, 28, 44, 45, 98, 101, 102, 213, 216

B

bacteria, 4, 6, 8, 12, 54, 55, 56, 58, 59, 61, 65, 71, 72, 73, 74, 75, 76, 77, 181, 185
beverages, vii, 1, 2, 18
biochemistry, 76, 224
biodiversity, 118
biological activities, 187
biosynthesis, 32, 123, 157, 175, 180, 185, 187
biotechnology, 187
blends, 23, 24, 202, 207, 217
Brazil, vi, viii, 1, 2, 5, 13, 16, 19, 43, 51, 83, 86, 119, 122, 129, 130, 131, 139, 140, 141, 143, 144, 145, 151, 152, 153, 155, 162, 197, 198, 199, 200, 201, 205, 206, 208, 209, 210, 211, 213, 216, 218, 219, 222, 223

C

C13 norisoprenoids, 134, 136, 137, 145, 159
C13 Norisoprenoids, 136, 137
C6 compounds, 134, 135
C9 norisoprenoid, 134, 135

cabernet sauvignon, v, 5, 16, 21, 22, 23, 24, 25, 26, 28, 29, 30, 31, 32, 33, 34, 36, 37, 38, 39, 43, 45, 46, 47, 51, 134, 157, 172, 188, 190, 202, 204, 210, 225
carmenère, vi, viii, 79, 80, 81, 82, 83, 84, 85, 86, 87, 88, 89, 90, 91, 92, 93, 94, 95, 96, 97, 99, 100, 101, 102, 103, 104, 105, 106, 107, 110
carotene, 137, 177
carotenoids, 133, 136, 137, 149, 161, 175, 177, 178
chemical, vii, viii, 2, 3, 4, 5, 7, 9, 11, 12, 13, 15, 16, 18, 36, 54, 56, 57, 58, 59, 62, 63, 64, 68, 71, 72, 74, 89, 98, 107, 111, 112, 116, 118, 133, 134, 139, 145, 146, 161, 164, 166, 179, 180, 184, 186, 187, 198, 202, 204, 210, 215, 219, 220, 223
chemical characteristics, 116, 222, 223
chemical composition, v, vii, 16, 51, 53, 56, 57, 72, 74, 112, 118, 164, 198, 204, 210, 218, 220
chemical degradation, 133
chemical properties, 13, 187
Chile, 24, 79, 80, 82, 83, 84, 85, 86, 87, 88, 95, 102, 103, 104, 106, 110, 113, 200, 205
chromatography, 95, 104, 111, 161
classification, 35, 36, 85, 105, 120, 193
climate, 12, 47, 75, 108, 118, 120, 128, 149, 155, 166, 198, 201, 202, 205, 206, 208, 209, 210, 218, 219, 225
commercial, 54, 55, 56, 57, 59, 61, 73, 75, 77, 88, 143, 149, 164, 173, 179, 191, 193, 195
condensation, 167, 168, 169, 170, 171
consumers, 2, 10, 14, 15, 43, 115, 153, 184, 186, 198
consumption, 47, 56, 98, 100, 174, 198
coronary artery disease, 174
coronary heart disease, 174, 191
correlation, 14, 50, 62, 165, 168, 176

cultivars, 5, 7, 12, 14, 15, 23, 30, 32, 47, 49, 180, 219
cultivation, 117, 158, 166
cultural practices, 175
culture, 56, 105, 181

D

degradation, 12, 34, 54, 58, 61, 67, 71, 81, 83, 123, 167, 177, 215
derivatives, 30, 31, 37, 89, 98, 99, 101, 124, 158, 165, 169, 170, 171, 180
detection, 47, 93, 94, 96, 104, 138, 158, 224
distribution, 32, 63, 99, 116, 120
diversity, 127, 186, 187, 198, 216
drying, 7, 8, 10, 12, 14, 16, 18, 19

E

environmental conditions, 98, 164
environmental factors, 55, 128, 198, 202
environmental temperatures, 156
enzymatic activity, 94
enzyme, 101, 150, 215
enzymes, 8, 136, 149, 180, 189, 194, 214, 215
ester, 62, 96, 184, 213
ethanol, vii, 4, 5, 6, 7, 12, 55, 94, 148, 168, 215
ethyl acetate, 70, 181, 213, 214
evolution, 50, 58, 65, 123, 124, 134, 169, 177, 178, 188, 190, 220, 221, 222, 225
exposure, 12, 32, 81, 87, 109, 117, 123, 149, 161, 178, 191
extraction, 6, 15, 19, 34, 44, 89, 90, 91, 107, 110, 153, 166, 168, 172, 179, 182, 188, 189, 203, 204, 214, 215, 216, 224, 225, 226
extracts, 26, 28, 29, 30, 33, 43, 45, 49, 158, 175

F

families, 62, 98, 133, 134, 145, 146, 169, 170, 171, 188, 195
fermentation, vii, 3, 4, 5, 6, 7, 8, 9, 18, 37, 51, 54, 55, 56, 57, 65, 67, 73, 74, 75, 76, 77, 85, 90, 91, 94, 96, 101, 107, 108, 110, 139, 142, 143, 146, 150, 161, 165, 167, 168, 169, 179, 180, 181, 182, 185, 188, 189, 190, 194, 196, 214, 215, 216, 217, 220, 226
fertility, 24, 80, 82, 84, 103
fertilization, 80, 82
flavanol, 164, 166, 169, 170, 171
flavonoids, 89, 98, 127, 129, 139, 143, 144, 165, 211, 212
flavonol, 99, 100, 101, 187
flavor, vii, 3, 11, 13, 55, 59, 62, 75, 92, 95, 109, 111, 117, 164, 165, 174, 179, 181, 185, 187, 192, 194, 201
flowers, 84, 86, 103, 213
food, 95, 97, 111, 165, 174, 186
formation, 9, 14, 60, 84, 89, 127, 159, 168, 169, 181, 184, 199, 210
France, 2, 21, 24, 34, 36, 46, 47, 48, 79, 80, 82, 86, 94, 102, 158, 160, 164, 174, 179, 197, 199, 200, 201, 213, 217
freezing, 101, 214, 215
fructose, 5, 57, 58, 65, 67, 68

G

geographical conditions, 198
glucose, 5, 15, 30, 57, 58, 65, 67, 68
glucoside, 15, 33, 37, 51, 81, 100, 113, 123, 124, 129, 132, 142, 145, 152, 172, 208, 211
grape, v, vi, vii, viii, 1, 2, 3, 4, 5, 6, 7, 8, 9, 10, 12, 13, 14, 15, 16, 18, 19, 22, 23, 24, 25, 26, 28, 29, 30, 32, 33, 43, 44, 47, 48,

49, 51, 56, 57, 59, 71, 74, 76, 77, 80, 81, 83, 86, 87, 88, 90, 91, 92, 94, 98, 99, 100, 103, 104, 105, 109, 111, 112, 113, 115, 116, 117, 118, 119, 120, 122, 123, 124, 125, 126, 127, 128, 129, 131, 132, 133, 134, 135, 136, 137, 138, 141, 145, 150, 151, 152, 153, 154, 155, 156, 157, 158, 159, 160, 164, 165, 166, 167, 168, 170, 171, 172, 175, 176, 179, 180, 181, 185, 186, 188, 189, 190, 191, 192, 194, 196, 197, 198, 200, 201, 202, 203, 204, 205, 207, 210, 214, 216, 217, 218, 219, 220, 221, 223, 224, 225, 226
grape and must composition, 120
grape berries, 81, 92, 94, 98, 99, 117, 118, 119, 120, 122, 127, 136, 153, 165, 214, 219
grape maturation, 117, 124, 125, 126, 128, 157
grape variety, vi, 23, 24, 28, 30, 33, 43, 49, 56, 57, 80, 92, 105, 115, 116, 117, 118, 123, 124, 125, 127, 128, 129, 131, 132, 133, 134, 135, 145, 150, 151, 153, 160, 166, 175, 179, 180, 186, 197, 200, 202, 207, 210, 217, 223, 226
growth, 6, 24, 32, 55, 59, 71, 77, 82

H

hue, 14, 30, 37, 118, 122, 141, 152, 167, 169
human, 23, 198, 202
humidity, 95, 121, 128
hybrid, 5, 7, 13, 18
hydrogen, 210
hydrolysis, 34, 62, 71, 80, 81, 96, 101, 149, 159, 172, 180

I

individual perception, 184
industry, 18, 151, 163, 167, 186
inoculation, 6, 7, 8, 54, 55, 56, 57, 58, 64, 65, 66, 67, 68, 69, 70, 72, 73, 74, 75, 77, 181
irrigation, 32, 49, 82, 199, 202

K

ketones, 69, 132, 180
kinetics, 56, 57, 75, 192

L

lactic acid, 6, 54, 55, 57, 60, 61, 63, 67, 68, 69, 73, 76, 77
late harvest, 80, 81, 85, 87, 88, 95
light, 6, 118, 123, 147, 149, 156, 157, 177, 178
liquid chromatography, 158
low temperatures, 7
luminosity, 79, 83, 87, 178, 215
lutein, 137, 177

M

malolactic fermentation, v, vii, 3, 6, 7, 8, 53, 54, 55, 57, 58, 64, 73, 74, 75, 76, 77, 90, 108, 141, 142, 143, 181, 194, 196
management, 5, 12, 32, 36, 80, 81, 86, 91, 92, 102, 136, 165, 175, 185, 199, 216, 222
mass spectrometry, 104, 111
maturation process, 2, 179
merlot, v, 5, 16, 21, 22, 23, 24, 25, 26, 28, 29, 30, 31, 32, 33, 34, 36, 37, 39, 42, 43, 46, 47, 48, 49, 50, 73, 80, 84, 86, 96, 99,

100, 101, 102, 113, 157, 188, 189, 202, 220, 224
metabolic pathways, 164, 185
metabolism, 50, 61, 76, 81, 92, 96, 98
metabolites, 59, 90, 164, 165, 179, 181, 185, 195, 204
methodology, 149, 184
methoxypyrazines, 79, 80, 81, 89, 92, 104, 105, 109, 196
microclimate, 32, 82
microorganisms, 4, 5, 56, 64, 74
mildew, 23, 24, 88, 117
molecular weight, 100, 127
molecules, 34, 89, 90, 92, 101, 164, 175, 204
monomeric anthocyanins, 37, 90, 127, 129, 143, 145, 151, 205, 208, 209, 211
monomeric flavonols, 127
monomers, 28, 43, 45, 98, 100, 128, 171
monoterpenic compounds, 134, 135, 146
multivariate analysis, 161
multivariate data analysis, 58

O

oligomeric procyanidins, 127
oligomers, 25, 27, 98, 126, 131, 171
oxidation, 34, 89, 91, 109, 124, 133, 167, 215
oxygen, 22, 44, 46, 52, 82, 89, 90, 91, 177
oxygen consumption, 22, 46
oxygen consumption rate, 22, 47

P

pH, 55, 57, 58, 59, 60, 63, 69, 85, 89, 94, 119, 122, 141, 142, 153, 169, 170, 176
phenolic, v, vi, vii, 2, 3, 6, 7, 10, 12, 13, 14, 15, 16, 19, 21, 25, 30, 34, 36, 38, 39, 47, 48, 50, 51, 71, 83, 85, 89, 90, 98, 99, 103, 107, 108, 109, 111, 112, 113, 115, 116, 119, 120, 127, 128, 129, 139, 142, 143, 144, 151, 152, 153, 155, 157, 158, 161, 162, 165, 166, 168, 171, 172, 173, 175, 185, 186, 187, 188, 189, 190, 191, 198, 202, 203, 204, 205, 208, 210, 211, 212, 214, 216, 217, 218, 220, 223, 224, 225, 226, 227
phenolic acids, vii, 11, 12, 89, 99, 143, 152, 171, 172, 175, 191
phenolic composition, v, 21, 22, 25, 34, 48, 50, 90, 98, 103, 107, 108, 112, 113, 115, 116, 127, 128, 139, 151, 153, 155, 173, 190, 191, 202, 204, 205, 208, 211, 212, 218, 220, 223, 225
phenolic compounds, vii, 2, 3, 10, 13, 15, 19, 22, 25, 30, 36, 38, 39, 51, 83, 98, 109, 111, 112, 119, 120, 127, 128, 129, 139, 142, 143, 144, 152, 158, 161, 165, 166, 168, 173, 185, 187, 198, 203, 204, 210, 212, 214, 216, 217, 218, 220, 225, 226
physicochemical properties, 18
pigmentation, 32, 50, 99, 100
polymeric flavanols, 128, 130, 131
polymerization, 26, 51, 100, 102, 131, 167, 172
polyphenols, 14, 22, 97, 98, 99, 100, 101, 112, 127, 157, 164, 165, 168, 174, 176, 187, 192, 215, 219, 225
Portugal, 33, 115, 116, 117, 118, 119, 120, 122, 123, 129, 130, 131, 139, 140, 143, 145, 154, 155, 156, 158, 160, 161, 162
precipitation, 120, 167, 210
principal component analysis, 63, 64
proanthocyanidin, 43, 49, 51, 102, 124, 125, 157, 158
procyanidin, 100, 125, 126, 128, 130, 131
prodelphinidin, 128, 172
producers, 1, 2, 22, 35, 36, 201, 202, 205

proteins, 28, 98, 100, 101, 102, 157, 165, 173
pruning, 24, 84, 86, 101, 176, 199, 207, 210, 222
pulp, 30, 88, 98, 117, 127, 166, 172

Q

quantification, 110, 142, 184
quercetin, 98, 99, 101, 172, 211

R

radiation, 121, 156, 157, 178, 199, 210
reactions, vii, 5, 46, 55, 89, 90, 98, 99, 124, 167
red wine, vi, vii, viii, 2, 5, 7, 12, 13, 14, 16, 17, 18, 19, 21, 22, 25, 35, 37, 43, 44, 46, 47, 49, 50, 51, 52, 55, 71, 74, 75, 79, 80, 81, 82, 89, 90, 92, 93, 94, 96, 97, 98, 99, 100, 101, 102, 104, 106, 107, 108, 109, 110, 111, 112, 113, 114, 115, 127,143, 144, 150, 153, 154, 158, 161, 167, 168, 172, 174, 181, 187, 188, 189, 190, 191, 192, 194, 195, 196, 201, 202, 203, 204, 205, 206, 207, 211, 212, 213, 214, 215, 216, 217, 218, 220, 221, 223, 224, 226
reducing sugars, 13
resveratrol, 14, 205, 211
ripening, 32, 49, 50, 81, 85, 87, 92, 103, 104, 112, 113, 124, 125, 156, 157, 158, 166, 167, 168, 171, 172, 176, 177, 178, 188, 190

S

secondary metabolism, 179, 180
seed, 14, 26, 28, 29, 30, 43, 49, 100, 101, 107, 113, 127, 128, 157, 168, 171, 188, 204
sensation, 22, 44, 91, 98, 100, 101, 143, 172, 174
sensory impact, 22, 220
sensory profile, vii, viii, 3, 7, 9, 11, 12, 14, 15, 16, 18, 51, 52, 54, 72, 73, 139, 140, 141, 181, 198, 202, 212, 213, 214, 218, 227
sensory properties, 16, 22, 56, 111, 164, 172, 179, 215, 221
sesquiterpenic compounds, 134, 136
showing, viii, 2, 3, 13, 56, 182, 198
skin, 25, 26, 29, 30, 31, 32, 33, 43, 49, 88, 89, 91, 98, 99, 100, 101, 107, 113, 124, 128, 137, 153, 158, 165, 168, 188, 200, 203, 226
solution, 6, 7, 50, 86, 89, 94, 168
South Africa, 24, 50, 73, 74, 75, 112, 157, 160, 161, 200, 223, 225, 226
South America, 35, 103, 218
Spain, 2, 24, 53, 54, 56, 57, 102, 106, 201, 205, 213
stability, 2, 15, 25, 30, 32, 36, 37, 55, 68, 89, 90, 91, 100, 108, 168, 170, 203, 223, 225
stabilization, 4, 6, 7, 8, 25, 89, 90, 98, 108, 220
storage, 51, 52, 89, 103, 108, 159, 217
structural characteristics, 130
structure, 25, 26, 27, 30, 31, 32, 45, 50, 71, 89, 103, 113, 190, 203, 204
sulfur, 4, 5, 7, 34, 90
sulfur dioxide, 4, 5, 7, 90
synthesis, 149, 174, 187, 195

T

tannins, 2, 22, 25, 26, 28, 29, 30, 34, 37, 45, 80, 81, 86, 89, 90, 98, 100, 101, 102, 111, 113, 131, 152, 168, 185, 190, 204, 207, 208, 209, 211, 212, 215, 224

techniques, 2, 3, 8, 14, 23, 34, 47, 89, 90, 91, 101, 168, 173, 189, 192, 196, 214, 217, 224, 225
technological maturity, 119, 120, 122
technology, viii, 4, 18, 116, 152, 176, 191, 192
temperature, 6, 7, 19, 47, 83, 84, 89, 101, 120, 121, 122, 128, 153, 156, 159, 178, 180, 214, 215, 225
tempranillo, v, viii, 43, 53, 54, 56, 57, 58, 59, 72, 73, 74, 76, 96, 108, 172, 202, 207, 224
terpenes, 12, 62, 71, 150, 153, 175
terpenoids, 118, 132, 138, 139, 151, 180
terroir, 36, 47, 81, 93, 105, 116, 118, 139, 153, 156, 165, 187, 202, 216, 219
Touriga Nacional, vi, viii, 50, 115, 116, 117, 118, 119, 120, 121, 122, 123, 124, 125, 126, 127, 128, 129, 130, 131, 132, 133, 134, 135, 137, 138, 139, 140, 141, 142, 143, 144, 145, 146, 147, 149, 150, 151, 152, 153, 154, 155, 157, 161, 162, 202
training, 86, 109, 137, 138, 139, 146, 150
treatment, 101, 108, 123, 179, 192, 215, 216, 224

U

ultrasound, 214, 216, 226
Uruguay, 35, 83, 163, 164, 169, 174, 179, 185, 188, 193, 194, 195, 200, 205, 221

V

variations, vii, viii, 3, 12, 32, 166, 182, 210
varieties, 14, 17, 21, 22, 23, 28, 30, 31, 32, 36, 37, 43, 47, 48, 49, 56, 71, 79, 81, 82, 84, 85, 86, 89, 90, 91, 94, 96, 98, 99, 100, 101, 103, 112, 113, 115, 116, 134, 136, 141, 155, 157, 158, 160, 165, 166, 167, 171, 172, 184, 185, 187, 190, 199, 202, 204, 207,210, 215, 217, 218, 225
véraison, 123, 124, 125, 126, 136, 149, 153
Vitis labrusca, v, 1, 2, 3, 4, 5, 6, 7, 10, 12, 13, 14, 15, 16, 17, 18, 19
Vitis vinifera L., vi, 47, 49, 50, 51, 104, 110, 112, 113, 116, 135, 137, 155, 156, 157, 158, 161, 187, 188, 190, 193, 197, 198, 199, 200, 202, 220, 221, 222
volatile composition, 51, 54, 60, 63, 69, 76, 108, 109, 145, 195, 223
volatile compounds, vii, viii, 5, 11, 58, 59, 62, 68, 71, 75, 92, 95, 132, 133, 134, 135, 145, 146, 147, 174, 180, 181, 182, 183, 213, 216, 227
volatile organic compounds, 180

W

water, vii, 12, 32, 82, 118, 148, 154, 168, 192, 199
winemaking, vi, vii, viii, 2, 3, 4, 7, 8, 9, 10, 11, 12, 13, 14, 15, 16, 17, 18, 23, 25, 43, 54, 55, 64, 65, 72, 80, 89, 90, 91, 92, 103, 107, 108, 112, 113, 114, 115, 116, 139, 151, 162, 165, 167, 168, 175, 179, 182, 186, 187, 189, 193, 194, 198, 202, 204, 214, 216, 220, 224, 225, 227
winemaking technology, vi, viii, 4, 115, 116, 139, 151, 214
wood, 2, 44, 46, 52, 89, 90, 182
wood products, 46
worldwide, vii, 1, 2, 3, 4, 10, 79, 80, 105, 197, 202, 214

Y

yeast, 3, 5, 6, 7, 13, 50, 54, 55, 56, 57, 65, 68, 69, 70, 71, 72, 74, 75, 91, 94, 96,

146, 165, 168, 169, 179, 180, 181, 186, 189, 194, 195, 218

yield, 8, 88, 103, 157, 167, 202